COMPREHENSIVE AUDITS OF RADIOTHERAPY PRACTICES: A TOOL FOR QUALITY IMPROVEMENT

COMPREHENSIVE AUDITS OF RADIOTHERAPY PRACTICES: A TOOL FOR QUALITY IMPROVEMENT

QUALITY ASSURANCE TEAM
FOR RADIATION ONCOLOGY (QUATRO)

INTERNATIONAL ATOMIC ENERGY AGENCY
VIENNA, 2007

Sales and Promotion, Publishing Section
International Atomic Energy Agency
Wagramer Strasse 5
P.O. Box 100
1400 Vienna, Austria
fax: +43 1 2600 29302
tel.: +43 1 2600 22417
email: sales.publications@iaea.org
http://www.iaea.org/books

Printed by the IAEA in Austria
October 2007
STI/PUB/1297

IAEA Library Cataloguing in Publication Data

Comprehensive audits of radiotherapy practices : a tool for quality improvement : Quality Assurance Team for Radiation Oncology (QUATRO) — Vienna : International Atomic Energy Agency, 2007.
p. ; 24 cm.
STI/PUB/1297
ISBN 92–0–103707–4
Includes bibliographical references.

1. Radiotherapy. — 2. Radiotherapy. — Equipment and supplies — Quality control. 3. Radiation — Safety measures. I. International Atomic Energy Agency.

IAEAL 07–00489

FOREWORD

As part of a comprehensive approach to quality assurance (QA) in the treatment of cancer by radiation, an independent external audit (peer review) is important to ensure adequate quality of practice and delivery of treatment. Quality audits can be of various types and at various levels, either reviewing critical parts of the radiotherapy process (partial audits) or assessing the whole process (comprehensive audits).

The IAEA has a long history of providing assistance for dosimetry (partial) audits in radiotherapy to its Member States. Together with the World Health Organization (WHO), it has operated postal audit programmes using thermoluminescence dosimetry (TLD) to verify the calibration of radiotherapy beams since 1969. Furthermore, it has developed a set of procedures for experts undertaking missions to radiotherapy hospitals in Member States for on-site review of dosimetry equipment, data and techniques, measurements, and training of local staff. This methodology involves dosimetry and medical radiation physics aspects of the radiotherapy process without entering into clinical areas.

The IAEA, through its technical cooperation programme, has received numerous requests from developing countries to perform comprehensive audits of radiotherapy programmes to assess the whole process, including aspects such as organization, infrastructure, and clinical and medical physics components. The objective of a comprehensive clinical audit is to review and evaluate the quality of all of the components of the practice of radiotherapy at an institution, including its professional competence, with a view to quality improvement. A multidisciplinary team, comprising a radiation oncologist, a medical physicist and a radiotherapy technologist, carries out the audit.

The present publication has been field tested by IAEA teams performing audits in radiotherapy programmes in hospitals in Africa, Asia, Europe and Latin America. Their comments, corrections and feedback have been taken into account, as well as the suggestions of the participants of the IAEA workshop Quality Assurance Team for Radiation Oncology (QUATRO), held in Vienna in May 2005. The QUATRO procedures have been endorsed by the European Federation of Organisations for Medical Physics, the European Society for Therapeutic Radiology and Oncology, and the International Organization for Medical Physics. The IAEA officer responsible for this publication was J. Iżewska of the Division of Human Health.

CONTENTS

1. INTRODUCTION

Independent external audits are a necessary part of a comprehensive quality assurance (QA) programme in radiation oncology [1–3]. Quality audits can be of various types and levels, either reviewing specific critical parts of the radiotherapy process (partial audits) or assessing the whole process (comprehensive audits).

The audits of radiation dose and other relevant medical physics procedures are well described in various IAEA and peer reviewed publications [4–7]. The IAEA through one of its technical cooperation programmes has received several requests from developing countries to perform more comprehensive audits of their radiotherapy services, either nationally or of individual institutions. The IAEA convened an advisory group, comprised of radiation oncologists and medical physicists, to devise guidelines for IAEA audit teams to initiate, perform and report on such comprehensive audits. The group was given the name Quality Assurance Team for Radiation Oncology (QUATRO).

The term audit, as used in this publication, is synonymous with an independent external evaluation, assessment or peer review. The audit methodology selected here places the emphasis on radiotherapy structure and process rather than on treatment outcome[1]. The value of an outcome oriented audit will be recognized, although it is not anticipated that the data from such audits will be accessible for this audit. The audit includes radiation oncology, medical physics and radiotherapy technology aspects of radiation treatment. This audit is intended to be comprehensive, but cannot be exhaustive as it is only a snapshot of a radiotherapy department at a specific point in time. Opportunities for improvement exist in all institutions.

To capture the actual level of competence of a department, the audit addresses simultaneously the issues of equipment, infrastructure and operation of clinical practice. A major part of the audit is patient oriented. Therefore, the structure of the present publication follows the path of patients from the diagnosis and the decision to treat, through treatment prescription, planning, treatment preparation and delivery, and then through the follow-up process. Clinical and medical physics procedures include radiation safety and patient protection when appropriate. Professional training programmes for radiation

[1] Treatment outcome depends on the multidisciplinary treatment of cancer patients; it seldom depends on a single modality and because of the timescale involved, it reflects the practice from 5–10 years ago, which is not necessarily related to the current practice. Finally, treatment outcome data are not always immediately available. To capture the treatment outcome data a follow-up audit after 3–5 years of the QUATRO audit would need to be organized.

oncologists, medical radiation physicists and radiation therapists (RTTs)[2] are given special attention.

The interpretation of the results of the audit is made against appropriate criteria of good radiotherapy practice (quality standards). As one example of such criteria, the IAEA has given a description of the design and implementation of a radiotherapy programme regarding clinical, medical physics, radiation protection and safety aspects [8].

The present publication presents guidelines for QUATRO audit teams. It contains checklists that may be considered helpful audit tools to be used flexibly by auditors, depending on the local situation. It does not represent one radiotherapy standard applicable to all visited departments. The objective is to provide a general audit methodology that can be applied in a range of economic settings. The audit includes an assessment of the ability of an institution to maintain the radiotherapy technology at the level corresponding to the best clinical practice in the specific economic setting (related to the ability of a country to sustain that technology).

1.1. BACKGROUND TO IAEA ACTIVITIES IN AUDITING

The IAEA has a long history of providing assistance for dosimetry audits in teletherapy in developing countries, for education and support of radiotherapy professionals, and for review of the radiotherapy process in a variety of situations. Teletherapy dosimetry audits have been widely performed by several national and international organizations for approximately 60% of the radiotherapy centres operating worldwide [4].

The IAEA, together with the World Health Organization (WHO), has performed thermoluminescence dosimetry (TLD) audits by mail to verify the calibration of teletherapy beams in radiation therapy departments (or hospitals) in developing countries since 1969 [5]. The programme aims at improving the accuracy and consistency of clinical dosimetry in radiotherapy hospitals worldwide. Over this period of 37 years, the IAEA/WHO TLD programme has verified the calibration of more than 6200 photon beams in approximately 1500 radiotherapy hospitals. Detailed follow-up procedures for TLD results outside the acceptance limits have been implemented since 1996, including on-site visits for which the IAEA has developed a standardized set of procedures

[2] The abbreviation RTT is used to describe a radiotherapy technology professional. Different terms for RTT are used in different countries, e.g. radiation therapist, therapy radiographer and radiation therapy technologist.

to aid the radiotherapy physics experts at hospitals in resolving dosimetric discrepancies [6]. These procedures include a review of the dosimetry equipment, data and techniques, verification measurements and training of local staff.

1.2. PURPOSE

The ultimate purpose of a QA audit is to assess the current situation and to improve the quality of the radiotherapy process at the reviewed institution or programme.

A comprehensive audit of a radiotherapy programme reviews and evaluates the quality of all the elements involved in radiation therapy, including staff, equipment and procedures, patient protection and safety, and overall performance of the radiotherapy department, as well as its interaction with external service providers. Possible gaps in technology, human resources and procedures will be identified so that the institutions affected will be able to document areas for improvement.

Radiotherapy centres operating at a high level of competence[3] would have the following characteristics:

(a) Be capable of delivering a sustainable radiotherapy service to international standards[4] (see the IAEA report [8] and Appendix I);
(b) Be capable of serving as a model for other radiotherapy centres in the country;
(c) Be capable of providing professional training for staff working in radiotherapy.

The high standard of radiotherapy services, once achieved, needs to be maintained over a long timescale to ensure the adequate sustainability of the centre's competence levels. A follow-up comprehensive audit would need to be organized after a period of three to five years through the IAEA, regional or national structures, or professional bodies[5], in order to demonstrate that the

[3] In Europe, the term 'centre of competence' is used; in other regions different terms are used, for example, in Africa, upon successful completion of the peer review process, the audited centre is nominated as a 'regional designated centre for training'.

[4] The standards achievable must be sustainable in the Member State's economic environment. Thus this will represent a value judgement of the auditors about the appropriateness of the infrastructure on-site and whether it is being used effectively.

[5] Such regional or national auditing structures remain to be developed.

standard of radiotherapy services delivered by such a centre continuously complies with the centre of competence criteria listed above.

Institutions in Member States may request an audit for the following purposes:

(a) For support in an application to become an accredited training centre for a region;
(b) To receive assistance to improve clinical practice;
(c) To strengthen their QA programme;
(d) To receive assistance to ensure that the requirements for patient protection are met;
(e) To serve as guidance for further departmental development;
(f) To document gaps in technology and practices in order to solicit funding from national authorities or other funding bodies, including the IAEA;
(g) To seek recognition as a centre of competence (see footnote on p. 6).

Such audits are not designed for:

(a) Regulatory purposes, i.e. the teams are not convened as an enforcing tool but solely as an impartial source of advice on quality improvement.
(b) Investigation of accidents or reportable medical events (misadministration). In the event of an investigation specifically into these aspects, a more focused audit is required.
(c) Assessment for entry into cooperative clinical research studies, as these are conducted by peers within the group involved in the study and are focused on the strict adherence of an institute to a single specified clinical protocol on a selected group of patients.

2. AUDIT STRUCTURE FOR QUATRO MISSIONS

2.1. REQUEST FOR AN AUDIT

Comprehensive audits in radiotherapy are voluntary. The request for an audit normally originates from the radiation oncology department to be audited. The administration of the institution or their national Ministry of Health may also request an audit. The head of the audited department should endorse it, in order to assure optimum cooperation, and to maximize the benefit of the audit.

The institution requesting an audit must have the basic equipment infrastructure to deliver good quality radiotherapy. This should include teletherapy and brachytherapy treatment machines supported by appropriate equipment for dosimetry, imaging and treatment planning, computers, and immobilization devices. Should the IAEA realize that these criteria are not met, it could offer guidance on how to achieve this basic level.

In order for the audit team to be chosen appropriately, as much information about the current status of the department and the reasons for the audit need to be received by the IAEA prior to the visit for the audit. It is the responsibility of the requesting institution to clearly formulate the purpose of the audit and to transmit this to the audit team.

2.2. COMPOSITION OF ON-SITE AUDIT TEAMS

The audit methodology is designed for execution by a multidisciplinary peer review panel, whose expertise is predominantly in radiotherapy. It is important that the members of the audit team include experts in all aspects of the programme to be audited. They must also be familiar with the audit methodology. Preferably, at least one member of the audit team should be able to interview members of the audited department in a language they understand.

The composition of the on-site visit team will depend on the scope, level and expected content of the audit visit, but will usually include as a minimum:

- A radiation oncologist;
- A radiotherapy physicist;
- An RTT;
- As appropriate, an engineer or other member with special competencies (e.g. in radiation protection) may be included.

2.3. PREPARATION FOR THE AUDIT

The success of an audit depends heavily on the thorough preparation of all parties involved, including the participating institution, the audit team and the sponsoring organization (IAEA).

2.3.1. Role of the institution

The institution's role is to:

(a) Formulate the objectives of the audit;
(b) Prepare data and relevant documentation to enable the auditors to complete their evaluation according to the format of this document (Sections 3–6);
(c) Provide material requested for any dosimetry audit;
(d) Identify and ensure participation of the individuals needed for the audit, although the audit team should be free to interview any staff member they deem appropriate;
(e) Inform the entire department and hospital management of the audit and its time frame;
(f) Provide the treatment records requested by the audit team, although the audit team should be free to review any of the records available;
(g) Provide any clinical records from outside the department deemed relevant to the cases reviewed.

2.3.2. Role of the auditors

Auditors are required to:

(a) Be familiar with the audit procedures, discuss their approach among themselves and allocate their responsibilities[6];
(b) Review the preparatory and background information prepared by the institution and that provided by the IAEA;
(c) Request additional information if necessary;
(d) Provide a comprehensive report about their visit.

2.3.3. Role of the IAEA

The role of the IAEA is to:

(a) Select an appropriate audit team;
(b) Inform the institution about the methodology (provide this document);

---

[6] Experts should consult the appendices to ensure that the terms commonly used are clearly specified for the audited department (e.g. treatment, session and patient).

(c) In collaboration with the requesting institution, prepare a clear outline of the objectives of the audit mission;
(d) Request all the necessary data from the institution (type of equipment, persons in charge, size of centre, type of centre, staffing and patient load);
(e) Brief the audit team, emphasizing the control on the dissemination of the report (Sections 2.6 and 2.7);
(f) Facilitate the introduction of the audit team to the institution;
(g) Review all prior interactions with the IAEA, including dosimetry audits, expert visits and special audits (e.g. recent TLD or other dosimetry audit results and expert reports). In cases in which no recent dosimetry audit has taken place, the IAEA will arrange one prior to the comprehensive audit.

2.4. GUIDING PRINCIPLES AND PROCEDURES OF AUDITS

Audits will evaluate the overall performance of the radiotherapy department. In the process, the team should obtain a comprehensive understanding of the total operation of the department. Auditors need to consider the interaction of the radiation oncology department with the other hospital departments involved in cancer management, such as gynaecology, surgical specialties and medical oncology, and with the hospital administration. Auditors must have free access to all relevant staff members to assess the free and efficient flow of information and cooperation between the different professionals involved.

Auditors must seek evidence for a patient orientated organization, with a culture of improvement through learning and openness to new technologies, and a culture of strong cooperation among staff members. An appropriate QA programme/system should be in place with the objective of continuous quality improvement.

If research is conducted, its integration into clinical practice must be judged; for example, auditors need to assess whether the publication level reflects the effort put into research.

The tasks to be performed during any clinical audit are described in Sections 2.4.1–2.4.3 below.

2.4.1. Entrance briefing

An entrance briefing is required to introduce the auditors to the various staff members and to discuss the methods, objectives and details of the audit.

The auditors should reassure the department that patient confidentiality will be respected.

2.4.2. Assessment

Both the infrastructure of the department and the overall radiotherapy programme will be audited. The infrastructure includes staffing, equipment and facilities. An examination of the radiotherapy programme from the initial introduction of the patient, evaluation and staging of the patient, treatment planning and delivery to follow-up will be carried out.

Checklists have been designed (Sections 3–6) to help auditors organize the audit programme and to ensure coverage of all relevant topics. The detailed programme of an audit depends on the reasons for the audit, and a selection of topics may be made from the full audit checklists, as appropriate. The tools available include:

(a) Staff interviews;
(b) A complete tour of the facility;
(c) A review and evaluation of procedures and all relevant documentation, including a review of treatment records;
(d) Practical measurements and other tests of the performance of local systems and procedures, where appropriate and relevant;
(e) Observation of practical implementation of working procedures.[7]

Aspects of the treatment process, which should have a coordinated input from clinicians, medical physicists and RTTs, should be audited by the whole team. Only specialized aspects of the treatment process will be audited by individual team members. A sign-off procedure by the auditing team, assuring the department of individual patient confidentiality may be required.

2.4.3. Exit briefing

It is essential that auditors present their preliminary feedback to the department. At the completion of the audit, the institution should convene appropriate members from all groups of the therapy team who were interviewed, for an interactive exit briefing. This will include time for questions, and

[7] Direct observation of patient treatment is part of the review of records. This may require consent from both the patient and the doctor.

should include a detailed and open discussion of all the findings of the experts. Initial recommendations could be made at this stage, if obvious.

Immediately after the audit, preliminary recommendations should be presented in written format. The institution should be encouraged to ask questions and make an initial response to the assessment. The steps intended by the institution to respond to the recommendations and improve the activities of the department should also be discussed and recorded.

When measurements have been performed as part of the audit, completed forms and calculations should be left with the institution (Section 5.1.4).

2.5. CONCLUSION OF THE AUDIT TEAM

Auditors are expected to comment on how well the institution has satisfied the criteria set out in the checklists. They will form and express an opinion regarding the appropriateness of the staffing in relation to the patient throughput. They are also expected to comment on type, quality and amount of equipment. An evaluation of quality of patient care will be given.

If the department wishes to expand to new areas of expertise, appropriate separate recommendations will be made.

Auditors may recommend whether a follow-up visit or internal audit is required. If the recipients of the audit report fail to implement recommendations and these are considered to be significant because of their potential impact on patient treatment outcomes, the recipients should be informed that they have the responsibility of notifying the regulatory authorities.

With respect to gaps in technology, infrastructure and procedures, the audit team may identify two levels of issues:

(1) Easily resolved areas. These may either require minor changes, which are easy to implement, or involve major changes that require modifications to infrastructure but are feasible for the department. These will be included in the detailed recommendations of the audit team.

(2) Major problems that cannot be resolved by the radiotherapy department without significant changes outside the hospital or without significant additional resources. The solution to these problems may require government action and, if so, the relevant recommendations need to be included in the audit report.

In some cases, the audit team may wish to recognize the centre as having been found to be in compliance with the IAEA criteria for a centre of competence.

2.6. THE AUDIT REPORT

The audit results are presented in the form of an audit report that consists of two parts, a summary report and a detailed report. The former will summarize the mission and its conclusion, while the latter will include the details of the audit, comments by the auditors, the audit conclusion and the recommendations, if any.

A useful audit report must contain conclusions formulated in an unambiguous way, with clear and practical recommendations.

To arrive at valid conclusions, the audit group should address a series of key topics and measurements, which will constitute the objective part of the report. These items will then be discussed in the broader perspective of the local radiotherapy organization and culture, in order to produce a comprehensive document describing the audited department. The report should be concise. A suggested structure includes:

(a) Objectives of the audit;
(b) A brief description of the audit activities;
(c) A description of the facility (infrastructure, workload, etc.);
(d) The findings and results of the visit (including checklists);
(e) Benchmarking if appropriate;
(f) Conclusions;
(g) Recommendations (to the institution, to the IAEA and to the government);
(h) Annexes.

It is important that the audit report mentions whether the site visit was welcomed or not. The degree of cooperation from the institution, department and various members of the radiotherapy team has a significant impact on the credibility of the final report. At all times, audit reports are confidential except for clearly designated recipients and the IAEA staff facilitating the audit.

It should be understood that while it is the responsibility of the IAEA experts to discuss shortfalls in the services of the audited institution, the audit does not necessarily commit the IAEA to rectify any deficiencies identified.

2.7. DISSEMINATION OF THE REPORT

The detailed audit report will only be sent to staff in responsible positions in the radiotherapy department, for example the head of the department, the chief medical physicist, the head RTT and other staff members whose role in the institution is significant to this audit.

In recent missions sponsored by the IAEA, it has been requested that a summary report be prepared by experts for dissemination to the relevant national authorities. Amongst these are the national TC Liaison Officer and the national permanent mission in Austria. This summary report will include a short description of the mission findings and its main conclusions. It should refer only to essential verifiable facts and exclude any value judgments.

Recommendations in the report will be directed to the institution and the national authorities, and to the IAEA. Recommendations to the IAEA should be confined to general statements, for example, the need for a follow-up visit. Only if the audit is performed in the context of a national Technical Co-operation Project, should specific IAEA interventions for training fellowships, expert missions or equipment be recommended.

3. INFRASTRUCTURE

3.1. AIMS OF A RADIOTHERAPY DEPARTMENT

The auditors will make an assessment of the adequacy of the objectives of the radiotherapy department in the context of national cancer care and of the degree to which the existing infrastructure is sufficient and properly used for addressing the objectives of the department.

3.1.1. Objectives of a radiotherapy department

The head of the radiotherapy department is responsible for answering the following questions about the department:

(a) Its role within the health care system;
(b) Its relationship with neighbouring oncology services (if any);
(c) Its relationship with other specialties within the hospital;
(d) Its role in teaching: undergraduate and/or postgraduate;
(e) Its role in research;
(f) Its current objectives (as they relate to quality, utilization of resources and institutional approach to patient care) and the documentation to support these objectives;

(g) Its financial structure and source of funding (State, private, etc.);
(h) Its vision and plans for the future.

3.1.2. Patient demographics

Auditors must familiarize themselves with the definition used to determine a 'new patient' and a 'new cancer' in order to assess patient numbers and statistics. A number of different conventions exist, some of which are addressed in Appendix II:

(a) Number of new cases (cancer or patients) per annum (Appendices II and III); Is information on new cases registered in a cancer registry?
(b) Types of cancer (primary sites and numbers);
(c) Stages of disease of the more common tumours;
(d) Source of information, for example, a cancer registry;
(e) Ratios of radical (curative) treatment to moderately high dose palliative therapy to palliative treatment;
(f) Fraction of cancer patients (of the total number in the catchment area) who come for radiotherapy, where the statistical data are available;
(g) Socioeconomic concerns with an impact on treatment[8] (payment required by hospital from patients, for example, medical insurance, private patient, government funded (free for patients) or co-payment).

3.2. STRUCTURE OF A RADIOTHERAPY DEPARTMENT

One of the important aspects of the audit is the assessment of staffing levels, the professional competence of the staff, the organization of work and the adequacy of the premises.

3.2.1. Personnel

Consideration of the following matters will help auditors gain an understanding of the appropriateness of staffing numbers in different professional groups and of their professional qualifications:

[8] The most common confounding factor is the proportion of the cost of therapy that is levied on the patients (and their families). In some societies, this will mitigate against the elderly or women receiving treatment.

(a) Number of radiation oncologists:
   (i) Professional qualifications (degrees, specializations, accreditations or fellowships);
   (ii) Additional responsibilities (e.g. chemotherapy or nuclear medicine).
(b) Number of medical physicists in radiotherapy, including clinically qualified radiotherapy medical physicists:
   (i) Professional qualifications (degrees, specializations, accreditations or fellowships);
   (ii) Additional responsibilities (e.g. diagnostics or radiation protection).
(c) Number of RTTs and their professional qualifications (degrees, specializations, accreditations or fellowships).
(d) Number of personnel assisting RTTs, for example, nurses.
(e) If there is no professional title in one or more of these professions, is there a local policy on education?
(f) What other members of staff (e.g., engineers, dosimetrists, nurses, social workers and psychologists) are there?
(g) Is there a programme for teaching junior medical staff (residents) and students? How many residents are there? How many medical students are there?
(h) Is teaching part of routine activity?
(i) Is research (basic and/or clinical) part of routine clinical activity?
(j) Are any staff allocated to clinical research?

The essential staffing levels are given in Appendix I and Ref. [8].

3.2.2. Departmental operation

The questions listed in this section will help auditors understand the organization of work in the department:

(a) What are the working hours (within the department) of the radiation oncologists, medical physicists and RTTs?
(b) What are the hours during which treatment is available at the department?
(c) How many days per week is the department in operation?
(d) Are emergency radiation services provided after hours?
(e) What is the minimum number of RTTs for each major item of equipment?
(f) What is the minimum number of radiation oncologists on duty during treatment hours?

(g) What is the minimum number of physicists on duty during treatment hours?

3.2.3. Premises

The physical layout of the department should be disclosed to the auditors in advance, prior to the audit. The following checklist may help the audit team to evaluate the adequacy of the premises in the context of the departmental objectives and operations:

(a) Location of the radiotherapy department relative to the main hospital (off-site, on-site, integrated into main building);
(b) Size and layout of the department:
   (i) Treatment rooms and control rooms;
   (ii) Examination rooms, changing rooms, consultation rooms, toilets and waiting rooms;
   (iii) Dosimetry and physics rooms, and laboratory space;
   (iv) Block cutting rooms (with appropriate ventilation) and storage rooms;
   (v) Secretarial areas and filing rooms.
(c) Proximity of radiotherapy department to teaching facilities, laboratories, etc.
(d) Is there access to additional sources of information about medical science, such as a library, research journals or the internet?
(e) Wards and number of beds (male, female and paediatric).
(f) Guest house?

3.2.4. Radiation therapy equipment

A full inventory should be made of all major equipment on-site, i.e. teletherapy equipment (status: functional, partially functional or redundant), brachytherapy equipment, imaging equipment, mould room and treatment planning. This would include non-functional and decommissioned equipment, which occupy useful space:

(a) Type, age and number of teletherapy machines.
(b) Type, age and number of brachytherapy units.
(c) Radioactive sources, storage facilities and radiation safety equipment.
(d) Available imaging equipment (including simulation).
(e) Available treatment planning equipment.
(f) Mould room equipment.

(g) View boxes, film processors and computerized networked imaging equipment.
(h) Immobilization devices.
(i) Patient alignment equipment, lasers, etc.
(j) Dosimetry equipment: phantoms, dosimeters, etc.
(k) Supporting equipment and spaces:
  (i) Secretarial areas, computers, printers, fax machines, typewriters and telephones;
  (ii) Access to filing rooms, storage and delivery of records (off-site or on-site);
  (iii) Patient information, waiting room chairs, wheelchairs and stretchers.
(l) Does the institution have an equipment replacement programme?
(m) Is there a calendar for preventive maintenance work?

A list of major equipment items relevant to a radiotherapy department is given in Appendix I.

3.3. COMMUNICATIONS

The relevant documentation illustrating the processes of dissemination of information throughout the radiotherapy programme should be prepared by the department and made available to auditors on-site:

(a) Record keeping and documentation (clinical and medical physics data).
(b) Across disciplines; access to hospital and physician records. Computer and fax equipment available. Adequacy of telephone communications.
(c) Horizontal communication (between staff members with the same function) and vertical communication (between senior and junior staff members).
(d) Between different areas of the radiotherapy process.
(e) Between staff on different shifts, when applicable.

3.4. QUALITY MANAGEMENT SYSTEM

The following functions, committees, training and equipment should be considered when reviewing the quality management aspects of the operations of a radiotherapy department:

(a) A QA committee;
(b) A quality manager (responsible for the programme);
(c) Frequency of quality review meetings and the written minutes of these;
(d) Meetings to discuss introduction of new techniques;
(e) Availability of quality control (QC) manuals;
(f) Directives on triggers and actions;
(g) Quality control procedures for each machine in department;
(h) Quality control records, including calibrations;
(i) Documentation of response to checks revealing equipment to be out-of-tolerance;
(j) Any other quality audits (internal or external);
(k) Training of personnel in the use of equipment.

3.5. RADIATION PROTECTION OF PATIENTS, STAFF AND THE GENERAL PUBLIC

Radiation protection and safety aspects of radiotherapy should be reviewed including the following items:

(a) Radiation protection committee;
(b) Manual of radiation protection;
(c) Record of personnel monitoring and feedback to staff;
(d) Radiation protection training and certification;
(e) Contingency plans (handling of incidents, deviations, etc.);
(f) Patient protection policy and procedures (justification and optimization).

3.6. WORKLOAD

3.6.1. Patient throughput on radiotherapy equipment

When assessing the quality of radiotherapy services, patient throughput on radiotherapy equipment is an important aspect to consider. The following data need to be made available to the auditors:

(a) The number of new cancer cases[9] or consultations of patients entering the department (This annual figure can be much larger than the number of radiotherapy treatments if the department integrates medical oncology and/or haematology).
(b) The number of new radiation therapy cases treated per annum in the department.
(c) The number of sessions/fractions[10] given monthly over a one year period by each teletherapy machine (T).
(d) The number of applications given annually by each brachytherapy machine (B)[11].
(e) The annual total number of computed tomography (CT) scans performed for planning purposes.
(f) The annual total number of simulations performed.
(g) The annual number of treatment plans generated by computer treatment planning.
(h) The relative proportion of simple, intermediate and complex treatments each machine delivers.
(i) The average treatment time on each machine.

Case accrual fluctuates during the year. The maximum daily figures give an indication of what the department can cope with when under pressure:

(j) Maximum number of fractions and fields in any one day on each therapy machine.

3.6.2. Statistics [12]

The following data should be considered when analysing the adequacy of the existing infrastructure in terms of human resources and equipment in the context of departmental operations:

[9] Appendix III provides details of annotations on the quantification of 'cancer cases'.

[10] Definitions are provided in Appendix II.

[11] Patients receiving both external beam radiotherapy and brachytherapy are thus recorded twice. Therefore, the number of individuals treated in a department is not simply the sum of T + B. Auditors should address this point unambiguously.

[12] Refer to Appendices II and III for the clarification of terms.

(a) The number of patients seen by a physician annually. It should be specified if the radiation oncologist also prescribes chemotherapy. Separate data for radiotherapy and chemotherapy should be given if appropriate.
(b) The number of patients per teletherapy machine annually.
(c) The number of treatment sessions per day.
(d) The average number of fractions per course of treatment.
(e) The number of courses of treatment per physicist annually.
(f) The number of treatment planning systems (TPSs) per physicist, RTT or dosimetrist (as applicable) annually.
(g) The number of courses of treatment per RTT annually.
(h) The number of treatment sessions or fractions per RTT annually.
(i) The number of RTTs per equipment item.

4. PATIENT RELATED PROCEDURES

Patient related procedures describing the clinical process are to be reviewed by the whole audit team, except for those sections where the expertise resides exclusively with radiation oncologists. In particular, Checklists 5 and 12–21 require input by medical physicists. Checklists 1, 2, 5, 10 and 11–23 are of interest to RTTs.

4.1. IDENTIFICATION OF PATIENTS

It is crucial that mechanisms be in place to ensure that the correct patient and the correct anatomical area of the patient be treated; otherwise, the risk of radiotherapy misadministration increases.

The precise system (e.g. an ID document and/or a photograph if economically feasible) shall depend upon national regulations regarding patient confidentiality. However, the audit team must ensure that an appropriate system is indeed in place and in use (Checklists 1 and 2).

CHECKLIST 1. IDENTIFICATION OF THE PATIENT AT THE START OF TREATMENT

Items to be reviewed by auditors	YES	NO	n.a.[a]
How is a patient identified at the start of treatment?			
Name	☐	☐	☐
Gender	☐	☐	☐
Address/Telephone number	☐	☐	☐
Age (date of birth, if known)	☐	☐	☐
National identification number (if any)	☐	☐	☐
Hospital identification number	☐	☐	☐
Departmental identification number	☐	☐	☐
File number	☐	☐	☐
Comments:			

[a] n.a.: not applicable.

CHECKLIST 2. IDENTIFICATION OF THE PATIENT ON A DAILY BASIS

Items to be reviewed by auditors	YES	NO	n.a.
How is a patient identified on a daily basis?			
Name	☐	☐	☐
One or more of the identification numbers in Checklist 1	☐	☐	☐
Photographic ID[a] (face)	☐	☐	☐
Photograph of the treatment site or field marks	☐	☐	☐
Anatomical sketch (diagram) showing location of treatment fields to be applied	☐	☐	☐
Other (e.g. code bar, etc.)	☐	☐	☐
Digital fields	☐	☐	☐
Are children handled differently from adults?	☐	☐	☐
Comments:			
How is patient confidentiality ensured?			

[a] ID: identity.

4.2. DIAGNOSIS AND STAGING

Investigations leading to tumour diagnosis and staging are necessary to deliver radiotherapy. Auditors will make an assessment of the degree to which the available infrastructure is used for patient diagnosis, staging and planning. The intent is to evaluate the presence and use of appropriate tools. Auditors may also consider recommendations on the introduction of cost effective additional investigations that may be justifiable.

Checklists 3–8 will document the existence and use of these tools:

CHECKLIST 3. CLINICAL RECORDS

Items to be reviewed by auditors	YES	NO	n.a.
Filing system	□	□	□
Clinical history	□	□	□
Physical examination	□	□	□
Comments:			

CHECKLIST 4. PATHOLOGY DOCUMENTATION

Items to be reviewed by auditors	YES	NO	n.a.
Location of pathology services:			
In the hospital	□	□	□
Outside the hospital	□	□	□
Is the pathology report in all patients' files?	□	□	□
The hospital's policy with regard to review of outside pathology services	□	□	□
Ability to obtain outside pathology consultations	□	□	□
Access to special stains, immunohistochemistry, hormonal receptors, etc.	□	□	□
Comments on the quality of service:			

CHECKLIST 5. ACCESS TO RADIOLOGICAL, ULTRASONOGRAPHIC AND NUCLEAR MEDICINE IMAGING

(Refer to Section 5.1.2, Checklist 24)

Items to be reviewed by auditors	YES	NO	n.a.
X rays	☐	☐	☐
Mammography	☐	☐	☐
Ultrasound	☐	☐	☐
Computed tomography	☐	☐	☐
Nuclear imaging (scintigraphy)	☐	☐	☐
Access to positron emission tomography (PET), etc.	☐	☐	☐
Access to magnetic resonance imaging (MRI)			
Delay (days) for diagnostic procedures	☐	☐	☐
Are significant radiological findings reported in the patient's chart?	☐	☐	☐
Comment on the quality of service (related to national resources), i.e. waiting times or any other impairment in access to staging procedures.			

CHECKLIST 6. ACCESS TO LABORATORY FACILITIES

Items to be reviewed by auditors	YES	NO	n.a.
Haematology	☐	☐	☐
Biochemistry	☐	☐	☐
Delay (days) to obtain results	☐	☐	☐
Access to immunology, genetics, etc.	☐	☐	☐
Are significant laboratory findings reported in the patient's folder?	☐	☐	☐
Comments on the quality of service (related to national resources):			

CHECKLIST 7. ENDOSCOPY PROCEDURES

Items to be reviewed by auditors	YES	NO	n.a.
Comments on endoscopy procedures:			
Are specialists and procedures available?	☐	☐	☐
Are there reports in patient charts?	☐	☐	☐

CHECKLIST 8. STAGING

Items to be reviewed by auditors	YES	NO	n.a.
Are patients staged and is this documented?	☐	☐	☐
Staging system used (e.g. TNM[a], AJCC[b], FIGO[c] and institutional):			
Consistency of documentation:			
Consistency of reporting of surgical staging, when appropriate	☐	☐	☐
Consistency of reporting of prior chemotherapy, when appropriate	☐	☐	☐
Performance status (WHO[d], Karnofsky or ECOG[e])	☐	☐	☐
Comments:			

[a] TNM: tumour, node, metastasis.
[b] AJCC: American Joint Committee on Cancer.
[c] FIGO: Fédération Internationale de Gynécologie et d'obstétrique.
[d] WHO: World Health Organization.
[e] ECOG: Eastern Cooperative Oncology Group.

4.3. INDICATIONS AND DECISION TO TREAT

Indications and decision to treat are based on clinical assessment and existing guidelines (Checklists 9–11). Any patient in the radiotherapy department must have had a treatment decision taken by a radiation oncologist.

CHECKLIST 9. MULTIDISCIPLINARY MEDICAL APPROACH

Items to be reviewed by auditors	YES	NO	n.a.
Are decisions to treat based upon meetings of multidisciplinary teams (tumour boards)?	☐	☐	☐
If yes, comment on meetings for:			
Every patient	☐	☐	☐
Specific types of tumour	☐	☐	☐
Frequency			
Meeting location (radiotherapy department or hospital)			
If not a multidisciplinary team, who generally refers the patient to the radiotherapy department (a general practitioner or a specialist)?			
Is the decision to treat inappropriately affected by outside factors? (economic, other specialties, etc.)	☐	☐	☐
Overall comments on multidisciplinary practice:			

CHECKLIST 10. PRACTICE GUIDELINES

Items to be reviewed by auditors	YES	NO	n.a.
Are written departmental protocols available for the most common clinical management situations?	☐	☐	☐
What is the source of guidelines followed by the department (hospital protocol manuals, national, international, textbooks or evidence based medicine)?			
Have clinical protocols been ratified by a departmental committee?	☐	☐	☐
How frequently are the treatment protocols reviewed?			
Are the tumour/site-specific protocols applied consistently within the department? (Are tumours at a particular site and stage treated in the same way?)	☐	☐	☐
Are regular meetings held to verify adherence to protocols?	☐	☐	☐
Is there coverage for absences of physicians from the department?	☐	☐	☐
Have all research protocols been ratified by an institutional ethics committee?	☐	☐	☐
Comments on the adequacy of guidelines and departmental policy:			

CHECKLIST 11. PATIENT INFORMATION AND CONSENT

Items to be reviewed by auditors	YES	NO	n.a.
Are the benefits and risks of radiation therapy explained to patients?	☐	☐	☐
How (leaflet, brochure and/or verbally)? Comments:			
Does a formal consent and agreement form exist on a patient file?	☐	☐	☐
Is there a protocol for the role of RTTs in the informed consent process? Comments:	☐	☐	☐

4.4. TREATMENT PREPARATION: INSTRUCTIONS FOR PLANNING

Preparation and planning phases must precede delivery of treatment and be completed in a precise and reproducible way. Checklists will assess the equipment and procedures used for localization, simulation and immobilization (Checklist 12), including mould room devices and procedures (Checklist 13).

CHECKLIST 12. LOCALIZATION, SIMULATION AND IMMOBILIZATION

(Refer to Section 5.1.2, Checklists 25 and 26)

Items to be reviewed by auditors	YES	NO	n.a.
Specify major equipment used for localization:			
Fluoroscopic simulator	☐	☐	☐
CT simulator	☐	☐	☐
CT dedicated to planning	☐	☐	☐
Diagnostic films taken on the treatment machine	☐	☐	☐
Portal films taken on the treatment machine	☐	☐	☐
Other (e.g. bone scan images)	☐	☐	☐
Film processors:	☐	☐	☐
Type			
Location relative to simulator?			
Are there view boxes near the simulator?	☐	☐	☐
Electronic imaging	☐	☐	☐
Comments on simulation:			
Comments on the integrity of geometric accuracy throughout the treatment preparation process and onwards (fiducial marks, coordinate system or lasers, flat couch on CT, etc.)			
Are localization/simulation resources used appropriately?	☐	☐	☐
Comments:			
Who is present during simulation/localization?			
Radiation oncologist	☐	☐	☐
Medical physicist	☐	☐	☐
RTT	☐	☐	☐
What is the role of the RTT/medical physicist/radiation oncologist, if present?			
Is a procedures manual available for simulation?	☐	☐	☐
Is there an exposure chart available (kVs and mAs)?	☐	☐	☐
Are X ray film geometric parameters available?	☐	☐	☐
Do the clinical tumour/site-specific protocols contain instructions for immobilization?	☐	☐	☐
Comments:			
Is there a field (skin) marking protocol?	☐	☐	☐
How are fields marked? How are marks maintained during treatment? How are marks documented for RTTs?			
Comments (tattoos):			

CHECKLIST 12. LOCALIZATION, SIMULATION AND IMMOBILIZATION
(Refer to Section 5.1.2, Checklists 25 and 26) (cont.)

Items to be reviewed by auditors	YES	NO	n.a.
RTT pre-treatment QC procedures (simulation, localization and planning):			
Simulation/portal film images: labels, date, field size, treatment parameters, signature of radiation oncologist.	☐	☐	☐
Comments:			
Is there adequate time for simulation procedures?	☐	☐	☐
Are procedures manuals available?	☐	☐	☐
Process for RTTs to review procedures manual. Describe.			
Contouring method (machine, wire, etc.). Who carries out contouring?	☐	☐	☐
Comments:			
Data transfer from imaging to planning:			
Manual transfer?	☐	☐	☐
Automatic transfer?	☐	☐	☐
Comments on data transfer:			

CHECKLIST 13. MOULD ROOM AND BEAM MODIFICATION DEVICES
(Refer to Section 5.1.2, Checklist 27)

Items to be reviewed by auditors	YES	NO	n.a.
Is a multileaf collimator (MLC) used?	☐	☐	☐
Are blocks standard?	☐	☐	☐
Is the inventory sufficient?	☐	☐	☐
Are standard blocks mounted?			
If blocks are to be mounted on a shadow tray, who mounts them?			
For unmounted blocks, how are blocks placed daily (i.e. template or skin marks)?			
Comment on standard blocks:			
Are blocks customized (individualized)?	☐	☐	☐
Has a mould room technician been appointed?	☐	☐	☐
Who designs the blocks?			
Who cuts the blocks?			
Are QA procedures performed on hot wire cutter (see also Section 5.1.2, Checklist 27)?	☐	☐	☐
Are customized blocks fixed to shadow trays?	☐	☐	☐
Is there a sufficient number of shadow trays for the clinical load?	☐	☐	☐
Is the melting point of the alloy used sufficiently low for the clinical throughput?	☐	☐	☐
How are blocks verified?			
Prior to treatment?	☐	☐	☐
With first portal film (image)?	☐	☐	☐
Comments on customized (individualized) block production and use:			
Comments on QA and the role of physicists in mould room procedures:			
Are there compensators?			

4.5. PRESCRIPTION AND PLANNING

This section describes auditing the process of teletherapy planning. The auditors will evaluate (Checklists 14 and 15):

(a) The interaction between different members of staff and whether they work well together as a functional unit;
(b) The means for ensuring the reproducibility of radiation administration;
(c) Quality assurance procedures.

CHECKLIST 14. TREATMENT PRESCRIPTION

(Refer to Section 5.1.2, Checklists 28 and 29)

Items to be reviewed by auditors	YES	NO	n.a.
Specify type of TPS			
Is there a procedures manual (treatment guidelines or protocols) for planning, including site-specific geometric arrangement of beams?	☐	☐	☐
Two dimensional (2-D) procedures (beam arrangements)	☐	☐	☐
Three dimensional procedures (organs at risk, definition of volumes, etc.)	☐	☐	☐
Proportions of manual, 2-D and 3-D treatments			
Are tumour volumes delineated?	☐	☐	☐
For curative (radical) patients?	☐	☐	☐
For palliative patients?	☐	☐	☐
Are the following target volumes used (see Reps 50 [9] and 62 [10] of the International Commission on Radiation Units and Measurements (ICRU)):			
Gross tumour volume (GTV)?	☐	☐	☐
Clinical target volume (CTV)?	☐	☐	☐
Planning target volume (PTV)?	☐	☐	☐
What are the recommended margins between CTV and PTV for each site/tumour technique?			
Comments:			
For which sites is planning optimization used?			
Does planning optimization involve:			
Definition of volumes?	☐	☐	☐
Definition of critical organs?	☐	☐	☐
Is the modality (photons or electrons) stipulated?	☐	☐	☐
Is the beam energy stipulated?	☐	☐	☐
Are the beam modifiers (e.g. wedges and blocks) stipulated?	☐	☐	☐
Is the patient position (e.g. supine or prone) stipulated?	☐	☐	☐
Is the dose per fraction stipulated?	☐	☐	☐
Is the total dose stipulated?	☐	☐	☐
Is the number of fractions stipulated?	☐	☐	☐
Is the total treatment time for schedules other than once daily five times per week stipulated?	☐	☐	☐

CHECKLIST 14. TREATMENT PRESCRIPTION
(Refer to Section 5.1.2, Checklists 28 and 29) (cont.)

Items to be reviewed by auditors	YES	NO	n.a.
Is the prescription signed by the radiation oncologist?	☐	☐	☐
Reporting system:			
ICRU?	☐	☐	☐
Other?	☐	☐	☐

CHECKLIST 15. TREATMENT PLANNING
(Refer to Section 5.1.2, Checklists 28 and 29)

Items to be reviewed by auditors	YES	NO	n.a.
Which treatment planning technique is used:			
Isocentric, source–axis distance (SAD)?	☐	☐	☐
Source–skin distance (SSD)?	☐	☐	☐
How are calculations performed:			
Manually?	☐	☐	☐
By computer?	☐	☐	☐
– 2-D TPS?	☐	☐	☐
– 2-D + TPS?	☐	☐	☐
– 3-D TPS?	☐	☐	☐
How many individuals check treatment calculations before first treatment?			
Are beam data in TPS:			
Generic?	☐	☐	☐
Specific?	☐	☐	☐
Are treatment machines uniquely identified in the TPS?	☐	☐	☐
Has the TPS the capacity to generate dose volume histograms (DVHs)?	☐	☐	☐
If so, are DVHs used by:			
Radiation oncologists?	☐	☐	☐
Medical physicists?	☐	☐	☐
RTTs?	☐	☐	☐
Other staff?	☐	☐	☐
Unused?	☐	☐	☐
Comments:			
Is there a policy on the maximum and minimum doses to the PTV?	☐	☐	☐

CHECKLIST 15. TREATMENT PLANNING
(Refer to Section 5.1.2, Checklists 28 and 29) (cont.)

Items to be reviewed by auditors	YES	NO	n.a.
Is treatment planning endorsed (signed) by the radiation oncologist?	☐	☐	☐
Is treatment planning endorsed (signed) by the medical physicist?	☐	☐	☐
Is treatment planning endorsed (signed) by an RTT (or other appropriate staff member)?	☐	☐	☐
What is the procedure if planning is not endorsed?			
Are there quality checks on treatment plans: protocols?	☐	☐	☐
Are there quality checks on dose calculations: protocols?	☐	☐	☐
Are there planning review meetings? If so, who are the participants? What is the frequency of these meetings?	☐	☐	☐
Comments on the quality of treatment planning:			

4.6. FROM PLANNING TO DELIVERY

This section describes the transition from planning to delivery (Checklist 16).

CHECKLIST 16. DATA TRANSFER FROM PLANNING TO DELIVERY

Items to be reviewed by auditors	YES	NO	n.a.
Are simulation images or virtual simulation images agreed prior to the course of treatment?	☐	☐	☐
Are data transferred from planning to delivery:			
Manually?	☐	☐	☐
Automatically?	☐	☐	☐
Through a record and verify system?	☐	☐	☐
Is the data transfer double checked? If so, what is the frequency of checks? Who is in charge?	☐	☐	☐
Comments (include QA, medical physics and RTT input):			

4.7. TREATMENT DELIVERY: TELETHERAPY

Auditors are encouraged to visit the different treatment units and explore the treatment delivery procedures directly on-site (Checklist 17). If the department treats children, auditors need to consider any necessary differences (general anaesthesia, immobilization, etc.).

CHECKLIST 17. TREATMENT DELIVERY PROCEDURES

(Refer to Section 5.1.2, Checklists 30 (megavoltage units) and 31 (orthovoltage units))

Items to be reviewed by auditors	YES	NO	n.a.
Is a patient log book kept at each individual machine?	☐	☐	☐
Computerized?	☐	☐	☐
By hand?	☐	☐	☐
Comments:			
How much time is allocated for the first treatment session?			
Are portal films obtained prior to or at the time of the first treatment?	☐	☐	☐
Is a physician present?	☐	☐	☐
Does the physician physically check the set-up?	☐	☐	☐
Does the physician check the portal film?	☐	☐	☐
Is a physicist present?	☐	☐	☐
– For all treatments?	☐	☐	☐
– For difficult set-up problems only?	☐	☐	☐
Is the physicist's presence mandatory?	☐	☐	☐
If not, is the physicist's presence an option?	☐	☐	☐
Is the patient psychologically prepared?	☐	☐	☐
If required, how are changes in the set-up managed?			
Patient set-up (positioning and immobilization):			
Skin marks?	☐	☐	☐
Tattoos?	☐	☐	☐
Are there any immobilization devices?	☐	☐	☐
Is there a diagram or photographs of the treatment position?	☐	☐	☐
Is a laser used for setting up?	☐	☐	☐
Is there a back-pointer?	☐	☐	☐
Is there portal verification?	☐	☐	☐
– If yes, when and how frequent?			
Is there a procedure for reviewing portal images?			
Who is responsible for any reviews?			

CHECKLIST 17. TREATMENT DELIVERY PROCEDURES
(Refer to Section 5.1.2, Checklists 30 (megavoltage units) and 31 (orthovoltage units)) (cont.)

Items to be reviewed by auditors	YES	NO	n.a.
Comments on organization of work at the treatment machine (cross-checks, etc.): How many RTTs are physically on-site for each treatment? What is the scheduled time allowed for each patient? What is the procedure used to ensure the correct patient is being treated with the correct fields and with the correct accessories? What procedure is used to deal with any side effects reported to RTTs or nurses?			
Is the beam set up (e.g. machine, modality, energy, aperture, gantry angle (GA), collimator angle (CA) and beam modifiers):			
Manually?	☐	☐	☐
Electronically?	☐	☐	☐
Electronically with manual verification?	☐	☐	☐
Comments:			
Who is authorised to override the treatment set-up? Comments:			
Monitoring units and treatment times:			
Is there an independent daily check of monitoring units?	☐	☐	☐
Is this cross-checked using a calculator on the first treatment day?	☐	☐	☐
Are routine checks made of treatment charts?	☐	☐	☐
If so, how often? By whom?			
How are patients monitored during exposure:			
By a video system?	☐	☐	☐
By an audio system?	☐	☐	☐
Others:			
In vivo dosimetry:			
Using thermoluminescence?	☐	☐	☐
With diodes?	☐	☐	☐
Other methods?			
Comment on the frequency of in vivo dosimetry and the patients on whom it is used.			

CHECKLIST 17. TREATMENT DELIVERY PROCEDURES
(Refer to Section 5.1.2, Checklists 30 (megavoltage units) and 31 (orthovoltage units)) (cont.)

Items to be reviewed by auditors	YES	NO	n.a.
Are all patients clinically reviewed during treatment? If so, how frequently? By whom (physician, nurse practitioner or RTT)?	☐	☐	☐
Does the facility have the infrastructure to manage combined chemotherapy and radiotherapy treatments?	☐	☐	☐
Comments:			
Is there a policy for handling interruptions in treatment?	☐	☐	☐
Is there a policy for handling patients who do not show up?	☐	☐	☐
Is there a policy for handling acute medical emergencies in the treatment room?	☐	☐	☐
Comments:			
Quality control procedures by RTT on treatment records and set-ups:			
Is there a policy on double checking of treatment set-ups?	☐	☐	☐
Is there a weekly QC of charts/records?	☐	☐	☐
Are set-up notes current and accurate?	☐	☐	☐
Are any field/dose parameter changes noted?	☐	☐	☐
Special instruction compliance (e.g. review films)?	☐	☐	☐
Are blood test compliance and results checked?	☐	☐	☐
Are any gap/separation changes noted?	☐	☐	☐
Are calculations redone?	☐	☐	☐
Are portal films/images retaken?	☐	☐	☐
Are portal films/images approved?	☐	☐	☐
Are daily treatment entries complete and signed?	☐	☐	☐
Are dose additions complete and correct?	☐	☐	☐
Have any new instructions from oncologist been checked?	☐	☐	☐
Are there records of teaching of nurses?	☐	☐	☐
Is there a procedure for patient care?	☐	☐	☐
Is patient condition documented together with follow-up measures?	☐	☐	☐
Is documentation complete?	☐	☐	☐
– Is storage and retrieval of patient documents satisfactory?	☐	☐	☐

CHECKLIST 17. TREATMENT DELIVERY PROCEDURES

(Refer to Section 5.1.2, Checklists 30 (megavoltage units) and 31 (orthovoltage units)) (cont.)

Items to be reviewed by auditors	YES	NO	n.a.
– What is recorded on the treatment sheet, how and by whom?	☐	☐	☐
– Is there a signature protocol?	☐	☐	☐
– Are there independent/double checks of the monitoring units delivered?	☐	☐	☐
– Are RTTs involved in patient reviews? If so, is this on a daily or weekly basis?	☐	☐	☐
Comments:			
Quality assurance procedures on clinical aspects of patient care and education:			
Is there a protocol on patient care?	☐	☐	☐
Is there a protocol on patient education (including psychosocial aspects)?	☐	☐	☐
Is there a Health and Safety protocol (including infection control)?	☐	☐	☐
Comments:			

4.8. DEVIATIONS IN RADIOTHERAPY ADMINISTRATION

A deviation in radiotherapy administration refers to any therapeutic treatment delivered to the wrong patient or the wrong tissue, or where a dose or dose fractionation differs substantially from the values prescribed; also any equipment fault, error, mishap or occurrence with the potential to cause patient exposures different from those intended (Checklist 18).

CHECKLIST 18. DEVIATIONS IN RADIOTHERAPY ADMINISTRATION
(Refer to Section 5.1.2, Checklist 35)

Items to be reviewed by auditors	YES	NO	n.a.
What would be regarded as an incident and what would not be regarded as an incident?			
Is the treating physician immediately notified of an incident?	☐	☐	☐
Is there a systematic reporting of incidents to a hospital committee?	☐	☐	☐
If so, is this verbal or written?	Verbal	Written	
Is a decision taken on the significance of the deviation?	☐	☐	☐
If so, is a significant deviation reported to the regulatory authorities?	☐	☐	☐
Have incidents been reported and, if so, how many?			
What is the RTT procedure for the reporting of error?			
Is there a system to enable anonymous reporting? Is there a 'no-blame' policy? Comment.			
What is the process for reviewing errors and 'near misses'?			
What is the policy on feedback?			
What is the policy on informing patients about incidents?			
What is the mechanism for corrective actions and how are RTTs involved? What is the mechanism for the implementation and monitoring of change?			

4.9. BRACHYTHERAPY FOR GYNAECOLOGICAL CANCER

This section audits the process of administration of brachytherapy to patients (Checklists 19–21). Gynaecological cancer is the most frequent indication for brachytherapy worldwide. If other brachytherapy activities are carried out regularly in a visited department, they should also be evaluated.

CHECKLIST 19. BRACHYTHERAPY INFRASTRUCTURE

(Refer to Section 5.1.2, Checklist 32)

Items to be reviewed by auditors	YES	NO	n.a.
Where is the brachytherapy treatment area relative to the teletherapy treatment area?			
Type of brachytherapy:			
Surface?	☐	☐	☐
Intraluminal?	☐	☐	☐
Intracavitary?	☐	☐	☐
Intraoperative?	☐	☐	☐
Interstitial?	☐	☐	☐
Intention to use brachytherapy:			
As a boost after external beam therapy?	☐	☐	☐
Alone?	☐	☐	☐
Intraoperatively?	☐	☐	☐
Mode of operation:			
Manual?	☐	☐	☐
Remote?	☐	☐	☐
Isotope and system used for intracavitary brachytherapy:			
Caesium-137 low dose radiation (LDR)?	☐	☐	☐
Caesium-137 medium dose radiation (MDR)?	☐	☐	☐
Iridium-192 high dose radiation (HDR)?	☐	☐	☐
Cobalt-60 HDR?	☐	☐	☐
Are radium devices still in use?	☐	☐	☐
Other (specify):			
Comments:			
Is there a verification system in place?	☐	☐	☐
X-ray?	☐	☐	☐
Endoscopy?	☐	☐	☐
Ultrasound?	☐	☐	☐
MRI?	☐	☐	☐
Is there a TPS in use?			
Application room design (space, shielding, etc.)			

CHECKLIST 20. BRACHYTHERAPY PROCEDURE
(Refer to Section 5.1.2, Checklists 32 and 33)

Items to be reviewed by auditors	YES	NO	n.a.
What types of applicators are used?			
Is there direct loading or after loading (manual or automatic)?	☐	☐	☐
Are there aseptic conditions for the insertion of applicators?	☐	☐	☐
Are applicators sterilized between uses?	☐	☐	☐
Are applicators for single use only?	☐	☐	☐
Comments:			
What type of anaesthesia/analgesia is generally used for:			
Cervix?			
Vagina?			
Others?			
Are ICRU guidelines for dose and prescriptions used?	☐	☐	☐
Is the application being carried out under the supervision of a radiation oncologist?	☐	☐	☐
For cervical cancer, what is the method of dose prescription/calculation:	Manual	Computer	
To point A?	☐	☐	☐
To point B?	☐	☐	☐
Other reference points?	☐	☐	☐
Rectum?	☐	☐	☐
Bladder?	☐	☐	☐
Other?	☐	☐	☐
In vivo dosimetry for cervix cancer treatment:	Yes	No	n.a.
Rectum?	☐	☐	☐
Bladder?	☐	☐	☐
Comments:			
Is insertion time pre-calculated or individually calculated?			
Transfer of TPS calculation in afterloading unit.			
Does the radiation oncologist validate the prescription?	☐	☐	☐
Is the physician present throughout the procedure?	☐	☐	☐
Who removes the applicators?			
Does the responsible physician see and sign the dose calculation?	☐	☐	☐

CHECKLIST 20. BRACHYTHERAPY PROCEDURE
(Refer to Section 5.1.2, Checklists 32 and 33) (cont.)

Items to be reviewed by auditors	YES	NO	n.a.
Does the responsible physicist see and sign the dose calculation?	☐	☐	☐
Are dose calculations cross-checked?	☐	☐	☐
What is the procedure for ensuring there is no source loss during treatment?			
If low dose rate non-automatic brachytherapy is employed, how are the medical and nursing staff informed of the time when the source is removed?			
What is the procedure for unloading (handling, transportation, storage of sources, etc.)?			
What training in safety (loading, unloading, handling, transportation, nursing and control of visitors) do staff receive?			
Are there emergency procedures?	☐	☐	☐
Are there repeated safety drills for HDR?	☐	☐	☐
Is there coordination in scheduling treatments by brachytherapy and teletherapy units?	☐	☐	☐

CHECKLIST 21. BRACHYTHERAPY REPORTING
(Refer to Section 5.1.2, Checklists 33 and 34)

Items to be reviewed by auditors			
How is the procedure recorded and reported?			
Which reporting system is used?	ICRU	Manchester	Other
	☐	☐	☐
	YES	NO	n.a.
Is the brachytherapy treatment integrated with external radiotherapy?	☐	☐	☐
If so, how is the dose calculated?			

4.10. TREATMENT SUMMARY (DOCUMENTATION)

This section refers to the recording and reporting of a treatment after its delivery (Checklist 22). In many countries there is a legal requirement for record keeping. Additionally, internal audit and clinical research require access to previous treatment data.

CHECKLIST 22. DOCUMENTATION OF THE TREATMENT SUMMARY

Items to be reviewed by auditors	YES	NO	n.a.
What happens to the treatment sheet after treatment?			
Is there a check by the physicist or other appropriate staff member (RTT)?	☐	☐	☐
Is there a treatment summary?	☐	☐	☐
How long are files kept? Where are they kept? Are they readily available?			
How long are treatment films kept? Where are they kept? Are they readily available?			
Is there a record of the treatment in the patient's (hospital) records?	☐	☐	☐
If yes, is there easy access to these records?	☐	☐	☐
Is a copy of treatment details sent to the referring physician?	☐	☐	☐
Is a copy of the treatment details given to the patient?	☐	☐	☐
Are cancer data communicated to a national/regional cancer registry?	☐	☐	☐

4.11. FOLLOW-UP

Follow-up of patients (Checklist 23) is essential to providing information with which to determine the effect of treatment (e.g. cancer control, side effects or misadministration). Follow-up is an important tool for internal and external audits. Auditors should evaluate the level of consistency of follow-up policy throughout the department.

4.12. REVIEW OF TYPICAL TREATMENTS

Examples of typical treatments of common cancer types are to be requested by the auditors for review and analysis, for example:

(a) Solitary bone metastasis in arms (non-weight-bearing bones);
(b) Multiple brain metastases;
(c) Radical treatment for common cancers (e.g. of cervix and lung);
(d) Breast cancer after conservative surgery;
(e) Brachytherapy, as appropriate.

CHECKLIST 23. PATIENT FOLLOW-UP

Items to be reviewed by auditors	YES	NO	n.a.
Do all radiotherapy patients receive a follow-up appointment after treatment?	☐	☐	☐
At what interval?			
Curative?	☐	☐	☐
Palliative?	☐	☐	☐
Is there a follow-up policy for the different cancers?	☐	☐	☐
Comments:			
For how long are patients followed up:			
One year?	☐	☐	☐
Two years?	☐	☐	☐
Five years?	☐	☐	☐
In excess of five years?	☐	☐	☐
Is the follow-up done in:			
The radiotherapy department?	☐	☐	☐
Elsewhere?	☐	☐	☐
Is the follow-up done by physicians other than radiation oncologists?	☐	☐	☐
Is the follow-up done by nurses or social workers?	☐	☐	☐
If follow-up is performed outside the radiotherapy department, are the reports on the outcome for patients available to the radiotherapy department?	☐	☐	☐
Are tumour control, failure of control and complications recorded at follow-up?	☐	☐	☐
Is radiation toxicity documented?	☐	☐	☐
Is radiation toxicity graded?	☐	☐	☐
Are the follow-up data analysed in terms of the above? By whom?	☐	☐	☐
Is there a policy of systematic review of serious complications?	☐	☐	☐
Comments:			

A representative number of cases of curative, palliative and post-operative treatments should be selected by auditors; the ratios of these types of treatments can be different in different departments.

The auditors should interpret these cases in relation to the funding of the department:

— Sufficient;
— Insufficient;
— Fee for service versus envelope per pathology;
— Fee per annum.

5. EQUIPMENT RELATED PROCEDURES

5.1. EQUIPMENT QUALITY ASSURANCE: ASPECTS RELATED TO MEDICAL PHYSICS

5.1.1. Introduction

The purpose of this part of the audit is to obtain an overview of the medical physics QA processes, procedures, documentation and records, as well as a sampling of the physics dosimetry data, to assess whether all appropriate physics aspects are covered and properly implemented. Auditors are again advised that the goal is to perform representative tests without being exhaustive.

The structure of an equipment related quality audit is similar to the overall audit structure and is mainly integrated with it, i.e. it is based on checklists, discussion with local personnel and observation. However, in addition, some limited measurements need to be carried out as part of the review of the dosimetry data set, along with sample checks of data consistency and some examples of clinical dose (and related) calculations for benchmark cases.

The data review, measurements and calculations are necessarily limited in scope by the time available. The measurements are only of basic parameters. The calculations are only for relatively simple situations. Therefore, the conclusions from the data evaluation are only valid within these limitations, i.e. of what it is possible to examine in this time.

If any significant discrepancies are observed, or if the data set appears not to be consistent, these observations should be recorded and discussed with the local physics personnel. The QUATRO physics expert may recommend that the IAEA identify an expert to visit the centre to perform more exhaustive tests as described in Refs [6] and [7].

The physics auditor is expected to be fully occupied with the structure and general process audit, along with the other auditors, in the normal time frame of the first three days of the group audit. Therefore, it is not possible to carry out measurements or the more detailed evaluation of data and calculations in that time frame. Instead, the physics auditor will normally expect to carry out these measurements on days four and five of the audit.

5.1.2. Quality assurance checklists for medical physics aspects

Equipment QC procedures and their documentation, and records, where appropriate, should be reviewed for all medical physics items.

The auditors should note who routinely performs the medical physics activities below: a resident medical physicist(s), a contracted medical physicist or other personnel to whom the duties have been delegated (Checklists 24–35).

CHECKLIST 24. IMAGING (INCLUDING X RAY, CT AND MRI UNITS)

Items to be reviewed by auditors	Comments
Specification of the equipment: Type Date of construction Date of installation	
Operation manual used	
Training of personnel for use of equipment	
Imaging procedures and involvement of medical physicist	
Quality assurance programme: Quality assurance manual Acceptance procedures Commissioning procedures Quality control programme (tests, frequencies, responsible persons, action levels and actions): – Warm-up procedure – Geometric accuracy, couch and lasers – Image quality (low and high contrast resolution, etc.) – Data display, data transfer and data manipulation – Accuracy and stability of CT data	
Incident log book	

CHECKLIST 24. IMAGING (INCLUDING X RAY, CT AND MRI UNITS) (cont.)

Items to be reviewed by auditors	Comments
Repair and maintenance programme: Log book Frequency Person in charge of repairs Procedure to accept repairs	
General condition of equipment and room:	

CHECKLIST 25. LOCALIZATION AND SIMULATION

Items to be reviewed by auditors	Comments
Specification of the equipment: Type Date of construction Date of installation	
Operation manual used	
Training of personnel for use of equipment	
Localization/simulation procedure and involvement of medical physicist:	
Quality assurance programme manual used	
Acceptance procedures	
Commissioning procedures	
Quality control programme (tests, frequencies, responsible persons, tolerance and action levels, and actions)	
Warm-up procedure	

CHECKLIST 25. LOCALIZATION AND SIMULATION (cont.)

Items to be reviewed by auditors	Comments
Mechanical and geometrical tests: Lasers Optical distance indicator (ODI) Central axis indicators Field size indicators Light and radiation field coincidence Angle indicators (GA and CA) Collimator axis of rotation: – Isocentre Gantry axis of rotation: – Isocentre Couch movements (vertical, lateral and rotational) Coincidence of simulator and couch isocentres Compatibility of couches and scales between simulator and treatment unit Field wires and contouring devices	
Image quality (dose rate, kVp and mAs calibration, high and low contrast resolution and film processing)	
Radiation protection	
Data transfer	
Incident log book	
Repair and maintenance programme: Log book Frequency Person in charge of repairs Procedure to accept repairs	
General condition of equipment and room:	

CHECKLIST 26. IMMOBILIZATION

Items to be reviewed by auditors	Comments
Role of physicist/RTT	
Acceptance, commissioning and QC of devices	
Dosimetry checks, when appropriate	
Communication	

CHECKLIST 27. MOULD ROOM AND BEAM MODIFICATION DEVICES

Items to be reviewed by auditors	Comments
Role of physicist/RTT	
Dosimetry checks, when appropriate	
Equipment and devices available	
Acceptance, commissioning and QC of devices	
Repair procedures, when appropriate	
Data transfer and verification	
Communication	

CHECKLIST 28. TREATMENT PLANNING

Items to be reviewed by auditors	Comments
Specification of the TPS: Type Date of installation/acceptance Latest upgrade	
Manual of operation/documentation of algorithms	
Training of personnel for use	
Quality assurance programme manual	
Acceptance procedures/reports	
Commissioning procedures/reports: Methods to obtain beam data Verification methodology	
Participation in external audits	
Control of consistency of TPS data with other departmental dosimetry data sets	
Quality control programme (tests, frequencies, responsible persons, tolerance and action levels, and actions): Test calculations/sample plans Checks of single field Checks of isodose distributions Reproduction of dose distribution for input data Monitoring of unit calculation Hardware input/output devices Data transfer	

CHECKLIST 28. TREATMENT PLANNING (cont.)

Items to be reviewed by auditors	Comments
Incident log book	
Upgrades of TPS: Log book Frequency Person in charge Procedure to accept changes	
Support from manufacturers (assistance in trouble shooting)	
Communication with manufacturers	
Links to user groups	
Is the TPS PC/workstation used for any other purpose than treatment planning (non-TPS software increases chances of corrupting the TPS files)?	

CHECKLIST 29. PATIENT DOSE CALCULATION PROCEDURES

Items to be reviewed by auditors	Comments
Responsibility for planning	
Manual of procedures	
Verification of introduction of new methods	
Request for planning and information provided	
Interactions with requesting physicians	
Comment on plan optimization methodology:	
Plan and chart checking methodology (tolerance and action levels)	
Storage and backup of plans	
Independent monitor unit (MU) calculation system and method	
Approval of plan	
Methodology for transfer of data for treatment delivery	
Procedures for plan changes during treatment	

CHECKLIST 30. TREATMENT DELIVERY: TELETHERAPY (COBALT UNITS AND LINEAR ACCELERATORS)

Items to be reviewed by auditors	Comments
Specification of equipment: Type Date of construction Date of installation	
Operation manual used	
Training of personnel for use	
Quality assurance programme manual	
Acceptance procedures[13]/reports	
Commissioning procedures/reports	
Participation in external audits	
Radiation safety surveys	
Quality control programme (tests, frequencies, responsible persons, tolerance and action levels, and actions)	
Warm-up procedures	
Safety tests: Door interlocks Radiation warning lights Area monitor (cobalt unit) Emergency on/off switches Manual means to shut off machine (cobalt unit) Exposure in room during 'beam-off' condition Collision avoidance Other safety interlocks	

[13] These should include an independent verification of the beam calibration.

CHECKLIST 30. TREATMENT DELIVERY: TELETHERAPY (COBALT UNITS AND LINEAR ACCELERATORS) (cont.)

Items to be reviewed by auditors	Comments
Mechanical and geometrical tests:	
Lasers	
Optical distance indicator (ODI)	
Central axis indicators	
Field size indicators	
Light and radiation field coincidence	
Angle indicators (GA and CA)	
Collimator axis of rotation:	
– Isocentre	
Gantry axis of rotation:	
– Isocentre	
Couch movements (vertical, lateral and rotational)	
– Isocentre	
Coincidence of collimator, gantry and couch isocentres	
Coincidence of mechanical and radiation isocentres	
Table top weight	
Beam dosimetry:	
Output constancy (daily tests)	
Dosimeter for daily tests	
– Calibration (certificate)	
– Constancy	
Beam calibration	
Field size factors	
Depth–dose dependence	
Beam uniformity	
Other systems (e.g. MLC)	
Clinical dosimetry:	
Beam dosimetry data:	
– Depth–dose data	
– Off-axis factors	
– Isodoses	
Monitor units/timer set calculations	
Wedge and tray factors	
Variation in SSDs	
Timer (co-unit: linearity and timer error)	
Monitor (linearity and proportionality)	
Gantry angle dependence:	
– Asymmetric jaws	
Special devices (e.g. stereotactic equipment)	

CHECKLIST 30. TREATMENT DELIVERY: TELETHERAPY (COBALT UNITS AND LINEAR ACCELERATORS) (cont.)

Items to be reviewed by auditors	Comments
Additional parameters for electron beams (e.g. cone ratios and gap factors)	
Special techniques, if any (e.g. total body irradiation (TBI))	
Advanced techniques, where appropriate (e.g. intensity modulated radiotherapy (IMRT))	
In vivo dosimetry: Equipment and methodology Calibration and QC Practical use Acceptance limits and corrective actions taken if results are outside these limits	
Portal imaging: Equipment and methodology Acceptance, commissioning and QC Practical use Acceptance limits and corrective actions taken if results are outside these limits	
Record and verify system, information network etc., as appropriate: Equipment and methodology Reports of acceptance tests Practical use Corrective actions taken if deviations occur	
Machine fault log book Procedure on occurrence of a fault	
Incident log book/reporting	
Repair and maintenance programme: Log book Frequency Person in charge of repairs Procedure to accept repairs	
General condition of equipment and room	

CHECKLIST 31. TREATMENT DELIVERY: TELETHERAPY (ORTHOVOLTAGE X RAYS)

Items to be reviewed by auditors	Comments
Specification of equipment: Type Date of construction Date of installation	
Operation manual used	
Training of personnel for use	
Quality assurance programme manual	
Acceptance procedures[14] and reports	
Commissioning procedures and reports	
Quality control programme (tests, frequencies, responsible persons, tolerance and action levels, and corrective actions taken)	
Safety tests: Door interlocks Radiation warning lights Other safety interlocks	
Mechanical and geometrical tests: Applicators Filters	
Beam dosimetry: Output constancy (daily checks) Dosimeter for daily checks: – Calibration (certificate) – Constancy Beam calibration Timer Half-value layer (HVL) check	
Clinical dosimetry: Beam dosimetry data: – Depth–dose data – Cross-beam distribution – Methods of treatment calculation	

[14] These should include independent verification of the beam calibration.

CHECKLIST 31. TREATMENT DELIVERY: TELETHERAPY (ORTHOVOLTAGE X RAYS) (cont.)

Items to be reviewed by auditors	Comments
Machine fault log book Procedure on occurrence of a fault	
Incident log book/reporting	
Repair and maintenance programme: Log book Frequency Person in charge of repairs Procedure to accept repairs	
General condition of equipment and room	

CHECKLIST 32. BRACHYTHERAPY

Items to be reviewed by auditors	Comments
Specification of equipment and systems: Type Date of construction Date of installation	
Operation manual used	
Training of personnel for use	
Quality assurance programme manual	
Acceptance procedures and reports: Source calibration: – Certificate – Traceability	
Commissioning procedures and reports	
Participation in external audits	
Quality control programme (tests, frequencies, responsible persons, tolerance and action levels, and corrective actions taken)	

CHECKLIST 32. BRACHYTHERAPY (cont.)

Items to be reviewed by auditors	Comments
Safety tests:	
Door interlocks	
Radiation warning lights and alarms	
Area monitor	
Portable survey meter	
Emergency on/off switches (LDR and HDR units)	
Emergency container and emergency kit for source handling	
Movable lead shields (manual LDRs)	
Exposure in room during ‘beam off’ condition	
Source dosimetry:	
Dosimeter (well-type chamber or equivalent)	
– Calibration (certificate)	
– Constancy	
Source calibration	
Uniformity of a batch of sources	
Uniformity of ‘linear’ activity	
Clinical dosimetry	
Imaging for source reconstruction:	
Accuracy of source positioning	
Coincidence of dummy and active sources	
Timer function	
Dose calculation algorithms and methods	
Other items:	
Source storage and disposal	
Transfer of sources	
Inventory of sources	
Source replacement policy	
Checking of contamination	
Source guides	
Mechanical integrity of applicators	
Machine fault log book	
Procedure on occurrence of a fault	
Incidents:	
Procedures for stuck or damaged sources	
Procedure for lost sources	
Log book	
Reporting	

CHECKLIST 32. BRACHYTHERAPY (cont.)

Items to be reviewed by auditors	Comments
Repair and maintenance programme: Log book Frequency Person in charge of repairs Procedure to accept repairs	
General condition of equipment and room:	

CHECKLIST 33. BRACHYTHERAPY TREATMENT PLANNING AND VERIFICATION

Items to be reviewed by auditors	Comments
Responsibility for planning	
Treatment planning equipment and methods	
Manual of procedures	
Verification of introduction of new methods	
Request for planning and information provided	
Imaging, localization and source positioning	
Interaction with the requesting physician	
Plan optimization methodology	
Plan and chart check methodology (tolerance and action levels)	
Independent calculation system and method	
Approval of plan	
Methodology for transfer of data to treatment delivery	
In vivo dosimetry if used	
Integration with teletherapy	

CHECKLIST 34. DOSIMETRY EQUIPMENT

Items to be reviewed by auditors	Comments
List of dosimetry equipment available (including barometers and thermometers)	
Operation manual used	
Acceptance and QC programmes (each item)	
Calibration of local standard ionization chamber, traceability and certification	
Calibration of field dosimeters	
Repair and maintenance programme	
General condition of equipment	

CHECKLIST 35. RADIATION PROTECTION AND SAFETY

Items to be reviewed by auditors	Comments
Responsibilities for radiation protection, for example: Persons identified Radiation safety officer appointed Responsibilities defined Awareness of these roles in department Radiation safety committee Radiation safety policy	
Licensing to conform to national requirements, for example: Licensing Authorization Accreditation requirements fulfilled: – For use of ionizing radiation – For facilities – For storage or disposal of radioactive material	

CHECKLIST 35. RADIATION PROTECTION AND SAFETY (cont.)

Items to be reviewed by auditors	Comments
Risk assessment and management, for example: Risk and hazard evaluations undertaken Range of possible incidents and accident scenarios considered Contingency planning for predictable events (instructions, corrective actions, investigations and reporting)	
Patient dose incidents and accidents (instructions, corrective actions, investigations and reporting)	
Consideration of radiation protection in planning of facilities and procedures	
Procedures for pregnant workers and pregnant patients	
Procedures for visitors, comforters and care-givers, and for discharge of patients, etc.	
Procedures for transport of sources to/from the centre and within the centre	
Classification and identification of areas (e.g. criteria and signs)	
Local rules for radiation protection in different areas (e.g. cobalt units, linear accelerators (linacs) and brachytherapy units)	
Local supervision of these rules	
Control of access	
Radiation protection equipment: Equipment available Acceptance, calibration and QC	
Radiation surveys, for example: What is done? Frequency of surveys Methods used Records made Corrective actions taken	
Practical procedures for personnel monitoring and investigation of significant doses Records made	

CHECKLIST 35. RADIATION PROTECTION AND SAFETY (cont.)

Items to be reviewed by auditors	Comments
Radioactive sources: Storage Security Inventory Handling Disposal Leak testing Records made	
Procedures for identification of authorized practitioners and operators Procedures for ensuring justification and optimization [11]	

5.1.3. Verification of consistency of dosimetry data and procedures

The suggested timescale for the more detailed evaluation or review of the consistency of the dosimetry data and procedures is immediately after the first three days of the audit. The timing of the dosimetry evaluation will have to fit in with the constraints on the physics auditor's time arising from the rest of the audit requirements and also those from access to the treatment equipment and local personnel. The audit pattern, as outlined below, proposes that the physics measurements be conducted between Day 1 and Day 4:

(a) Days 1–3: Common activities with other members of the audit team.
(b) Evening of Day 3: Measurements on at least one teletherapy machine. If a multimode linac is present, then measurements will be carried out on at least one X ray beam and at least one electron beam. If it is not a multimodal linac that is in use, but an X-rays-only type, then measurements will be undertaken on at least that machine and, finally, if no linacs are in use, measurements will be carried out on at least one cobalt unit.
(c) Day 4 morning: Measurements on at least some brachytherapy source systems.
(d) Day 4 afternoon: Measurement data analysis; a more detailed evaluation of the dosimetric data sets available, both manual and in the TPS; and some dose and related calculations for a selection of benchmark cases.
(e) Day 4 evening, if required: If any inconsistencies or problems are observed from previous measurements, data evaluation or calculations,

access to the treatment machines may be required to carry out any additional measurements the auditor may deem necessary to resolve or further investigate these issues.

(f) Day 5: Discussion between the physics auditor and the local physics personnel on the dosimetry data and processes, as part of the overall feedback to the department.

The suggested measurements are relatively simple and are for basic parameters only. The calculations are also for relatively straightforward situations. However, the depth of audit in the given circumstances relies on the judgement of the auditor.

The radiotherapy department needs to know in advance about the measurement programme but to adapt an attitude of flexibility in preparing for the visit of the auditors.

The recommended tests are to remain as given in this book and to be performed at the judgement of the auditor.

Comprehensive audits of electron beams are very time consuming, and so engaging in such audits relies on the judgement of the auditors in determining the depth of the audit.

5.1.3.1. Dosimetry for external beam radiotherapy

(a) Basic safety tests

The auditors should perform the following checks to ensure the radiation safety of working conditions before conducting any tests on the treatment unit:

(1) Door interlock operational;
(2) Radiation warning light operational;
(3) Exposure within treatment room with the treatment unit in ‘beam-off’ condition.

The auditor shall wear a personnel radiation monitoring device and, if available, a radiation survey meter with an active alarm option.

(b) Mechanical tests

A few basic geometrical tests are necessary to ensure proper set-up conditions for the calibration of the radiotherapy unit, as well as the positioning of patients for daily treatments. Any differences should be noted between the auditor's measurements and the institution's values. The minimum tests involve the following checks of lasers, ODIs and field sizes:

(1) Lasers: The congruence of the lateral lasers and the isocentre horizontal plane, 20 cm on either side of the isocentre, at the nominal treatment distance.
(2) Optical distance indicator (ODI): The congruence of the ODI and the mechanical isocentre; the ODI at –10 cm, and +10 cm from the mechanical isocentre.
(3) Field size indicator: The field size indicator compared with the light field at the nominal treatment distance for three field sizes (5 cm × 5 cm, 10 cm × 10 cm and 20 cm × 20 cm).

Once the auditor has verified these geometric parameters, they should proceed to make the dosimetry measurements outlined in the following section.

(c) Dosimetry calibrations and measurements

Before performing the beam output calibration, it is necessary for the auditor to perform a comparison between the institution's and the auditor's barometer and thermometer.

The local medical physicist should calibrate, under the observation of the auditor, the beam output according to the local institution's standard procedure for at least one photon beam and at least one electron beam if available (more if time allows). The auditor should follow carefully the whole procedure step by step and try to understand the local procedure completely.

The auditor will perform a beam output calibration for each of the above beams according to the IAEA code of practice described in Ref. [12] and compare the measured output with the institution's specification.

5.1.3.2. Clinical dosimetry

At this stage, the auditor should have knowledge of the clinical techniques routinely used at the institution. The auditor should therefore concentrate their efforts on the relevant clinical dosimetry data.

Some of the items described in the following have already appeared in the checklists; however, they are repeated here for completeness. It is assumed that during the course of the normal procedural audit the auditor may not have performed a full data evaluation and review; therefore, the intention here is to perform a more detailed evaluation.

(a) Basic dosimetry data

The auditor will review the beam data tables available, determine if the data are measured or based on published data and obtain copies of appropriate data (if possible).

The auditor will review and evaluate the consistency of the basic beam dosimetry data used by the institution by comparison with expected standard data. They will ascertain how the basic dosimetry data set is used in the TPS or the in-house software.

(b) Monitor units or time set calculation

The auditor will evaluate the institution's method used routinely to calculate the number of monitor units or time set for patient treatments. This needs to be done for all photon beams and at least one electron beam per linac. For this the local physicist must be requested to determine monitor units or time set for the clinical dosimetry tests as described below. In addition, the auditor will independently calculate the monitor units/time set for the same standard dosimetry tests using the output value that they have measured and the standard data supplied. The auditor's results will be compared with those determined by the institution.

The standard clinical tests will be performed for a simple water phantom treated with a single field. The monitor units or time set need to be calculated to deliver 2 Gy at the various points of interest. The following set-ups are recommended:

(1) Photon beams:

(i) Field size 10 cm × 10 cm, depth 10 cm, with and without a steep clinically used wedge;
(ii) Field size 7 cm × 20 cm, depth 10 cm, with and without the same wedge.

(2) Electron beams:

(i) Field size 10 cm × 10 cm, depth of calibration;
(ii) Field size: large applicator, with rectangular cut-out, depth of calibration.

If blocks are used at the institution, the auditor and the local physicist will calculate the monitor units or time set for a typical clinical blocked field used at the institution.

5.1.3.3. *External beam treatment planning system*

The auditor will perform a set of tests to verify the performance of the TPS:

(1) Confirm that the field sizes on printouts and the entered field sizes match;
(2) Compare a sample of dosimetry data with the expected data (at least including open and wedged fields);
(3) Observe and question the process to produce plans and calculations for at least one or two normal clinical cases.

5.1.3.4. *Brachytherapy*

(a) Basic safety tests

Before conducting any tests on the brachytherapy unit, auditors need to check the availability and functionality of the following equipment to ensure the radiation safety of the working conditions:

(1) Door interlocks, warning lights and alarms (in particular for HDR afterloading units);
(2) An area radiation monitor that is safe against power failure and a portable survey meter;
(3) The emergency container and emergency kit for source handling in the case of a failure of the source to retract into its storage container (HDR afterloading units);
(4) Movable lead shields (manual LDR source handling).

In addition, the auditors need to check the exposure within the room with the treatment unit in the 'source off' condition.

The auditor shall wear a personnel radiation monitoring device and (for manual LDR source handling) a finger dosimeter.

(b) Check of source calibration

The local medical physicist will check, under the observation of the auditor, the source calibration (in terms of the reference air kerma rate) for at least one source of at least a selection of activities, according to the local institution's standard procedure for brachytherapy units (remote afterloading), or a sample of individual sources (wires or seeds for manual afterloading). The auditor will carefully follow the local procedure step by step, trying to understand it completely.

The auditor will perform, respectively, checks of the source calibrations according to the guidelines given in Ref. [13] using a calibrated well-type ionization chamber.

(c) Clinical dosimetry

At this stage the auditor will have knowledge of the clinical techniques routinely used for brachytherapy at the institution. Therefore, the auditor now needs to concentrate their efforts on the relevant clinical dosimetry characteristics.

The auditor will perform a set of tests to verify the performance of the brachytherapy TPS or the planning calculation method:

(1) Compare a sample of dosimetry data with the expected data for standard brachytherapy applications;
(2) Observe and question the process to produce plans and calculations for at least one normal clinical case (including reconstruction, source distribution and time).

5.1.4. Exit interview and the end-of-mission report

As their contribution to the exit interview and the end-of-mission report, the medical physics expert should prepare a preliminary report of the review of medical physics procedures. The expert should leave a copy of their signed and dated measurements, calculations, a report of the results and a copy of Ref. [12] (if not available at the institution) to the local physicist. These data and information will provide the institution's physicist with a set of independently measured reference data that can be used later to compare their own measurements for possible future dosimetry changes. Any records left at the institution should be clearly marked 'preliminary'.

The expert may be required to address, to the radiation oncologist, any important changes recommended in dosimetry practices that might have an impact on the clinical outcome of patient treatments.

The end-of-mission report to the IAEA should contain the following data and information:

(a) A summary of the tests and measurements performed by the expert;
(b) Results of the measurements;
(c) Results of clinical dosimetry;

(d) Analysis of the results of the measurements;
(e) Recommendations to the institution: General and specific;
(f) Recommendation to the IAEA/WHO TLD postal dose audit programme.

The relevant forms, spreadsheets and worksheets [6] should be properly dated and signed.

5.2. EQUIPMENT QUALITY ASSURANCE: ASPECTS RELATED TO RADIATION THERAPISTS

5.2.1. Introduction

The RTT audit structure is integrated into the overall audit, which is based on checklists, discussion with local personnel and observation. Infrastructure and both patient related and equipment related procedures require the input of an RTT auditor, as appropriate. The RTT auditor is expected to be fully involved during the general audit, along with the other auditors.

During the first three days of the audit, the RTT auditor will gain an insight into the management structure and organizational relationships of the department and the level of responsibility expected of the RTT in the specific context of this department. In addition, the RTT auditor must spend time in a clinical setting with the RTTs. During this time, the RTT auditor is advised to observe the normal working conditions of the RTTs and to discuss with them the topics identified in the RTT checklists in more detail.

The purpose of this part of the audit is to obtain an overview of the role of the RTTs within the multidisciplinary team in radiation oncology with regard to the preparation and delivery of radiotherapy, with a special focus on equipment QA.

5.2.2. Quality assurance checklists: Aspects related to radiation therapists

Quality assurance procedures and practices, as well as QC protocols and records, need to be reviewed by the auditor for all items involving the practice of RTTs. Checklist 36 will help the auditor to review the quality of the RTT infrastructure. Quality control procedures for equipment are listed in Checklist 37.

CHECKLIST 36. RADIOTHERAPIST QUALITY ASSURANCE INFRASTRUCTURE

Items to be reviewed by the RTT auditor	YES	NO	n.a.
Quality assurance infrastructure:			
Is the role of RTTs articulated (in their job description)?	☐	☐	☐
Is there a reporting structure?	☐	☐	☐
Are RTTs autonomous?	☐	☐	☐
Number of days per week department is in operation: Number of hours per day department is in operation: Normal working hours per day/per week:			
Is there a shift system and protocol for changeovers?	☐	☐	☐
Number of RTTs per teletherapy unit:			
– Superficial X ray unit:			
– Orthovoltage X ray unit:			
– Cobalt-60 unit:			
– Single energy X ray linac:			
– Multienergy X ray plus electron linac:			
– Simulator:			
– Others:			
Average number of patients/fractions treated per teletherapy unit/day:			
Are RTTs knowledgeable about treatment protocols?	☐	☐	☐
Is there an orientation programme for new RTTs?	☐	☐	☐
Do RTTs participate in equipment selection?	☐	☐	☐
Do RTTs participate in training provided by vendors?	☐	☐	☐
Comments:			
Radiation safety:			
Are RTTs familiar with the radiation safety protocol for patients, staff and public?	☐	☐	☐
Comments:			
Departmental policies and procedures for QA:			
Do RTTs contribute to QA procedures?	☐	☐	☐
Is there a procedure for RTTs to question deviations?	☐	☐	☐
Comments:			

CHECKLIST 37. QUALITY CONTROL CHECKLIST FOR RTTs

Items to be reviewed by the RTT auditor	YES	NO	n.a.
Quality control procedures on imaging units:			
Are there QA procedures on film processing equipment?	☐	☐	☐
Is there a policy for radiation safety in simulator: call-out 'screening' and checking lead aprons?	☐	☐	☐
Is the consistency of all table tops, laser lights, field sizes and gantries checked?	☐	☐	☐
Are door interlocks checked?	☐	☐	☐
Are room monitors checked?	☐	☐	☐
Comments:			
Quality control of radiation oncology laboratory (mould room):			
Is there a procedure for checking the construction of immobilization/positioning devices?	☐	☐	☐
Is there a procedure for checking the construction of shielding devices?	☐	☐	☐
Are there QC procedures on:			
– Remote afterloading brachytherapy units?	☐	☐	☐
– Manual afterloading intracavitary/interstitial sources and surface applicators?	☐	☐	☐
Comments:			
Quality control procedures on treatment units:			
Is the consistency of all table tops, laser lights and field sizes checked?	☐	☐	☐
Are door interlocks checked?	☐	☐	☐
Are room monitors checked?	☐	☐	☐
Are quality checks carried out on accessory equipment at the point of use?	☐	☐	☐
Are quality checks carried out on immobilization devices: storage and replacement?	☐	☐	☐
Comments:			

6. TRAINING PROGRAMMES

The auditors are required to assess if there are professional education and training programmes for any of the professional classes of personnel, i.e. radiation oncologists, radiotherapy medical physicists and RTTs.

6.1. ACADEMIC PROGRAMME

The following questions should be clarified about any academic programmes:

(a) What is the minimum entry requirement for training?
(b) Is the education university based?
(c) What is the duration of training?
(d) Is there an internal audit process?
(e) Is there an external audit process?
(f) What qualification is required for the academic staff?
(g) List the personnel available for teaching of basic sciences.
(h) List the personnel available for teaching clinical oncologists.
(i) List the personnel available for teaching of medical physicists.
(j) List the personnel available for teaching RTTs.
(k) Provide the written programme for training in each discipline, if such programmes exist.

6.2. CLINICAL PROGRAMME

The following questions should be clarified about any clinical programme:

(a) Is there a training programme in the department?
(b) Is the department accredited for training?
(c) Who is the accrediting body?
(d) Is accreditation recognized locally, nationally or internationally?
(e) What kind of training takes place — internal or external?
(f) What is the duration of training?
(g) Is training done only in a single department?
(h) Is more than one centre included in the clinical training programme?
(i) Is there easy access to external training at national or international level?

(j) Is external training voluntary or compulsory?
(k) Is funding available to support external training?
(l) What is the ratio of trainees to qualified staff?
(m) Are there accredited trainers in the department?
(n) What is the ratio of students to trainers for each discipline?
(o) How is the training assessed/recorded? Provide details.
(p) Are all staff involved in clinical training?
(q) List personnel available for teaching basic sciences.
(r) List personnel available for teaching clinical oncology.
(s) List personnel available for teaching medical physics.
(t) List personnel available for teaching RTTs.
(u) Provide the written programme for each discipline, if such programmes exist.

6.3. RESEARCH

The following questions should be clarified about any research carried out:

(a) Is research included as part of the primary academic programme?
(b) Is clinical research supported in the department?
(c) Are all staff disciplines in the department involved in clinical research?
(d) Is ethical approval required?
(e) Is there a departmental research committee?
(f) Are all disciplines represented on the departmental research committee?
(g) Is there an ethics committee in the hospital?
(h) Is ethical approval required from more than one committee, for example, from a university as well?

6.4. PROFESSIONAL ACCREDITATION

The following questions should be clarified about professional accreditation:

(a) What level of certification is awarded?
(b) Who is the awarding academic body?
(c) Is there professional accreditation?
(d) Who is the accrediting professional body?
(e) Is a minimum level of competence defined?

(f) Is the qualification recognized internationally?
(g) Is there a process for recognition of non-national qualifications?
(h) Is there a professional title?

6.5. CONTINUOUS PROFESSIONAL EDUCATION

The following questions should be clarified about continuous professional education (CPE):

(a) Is there a CPE programme in the department?
(b) Is the CPE programme compulsory?
(c) Is there a national policy on CPE?
(d) Who facilitates CPE?
(e) Is CPE necessary for continuing registration?
(f) Are external components recognized?
(g) Does the department provide CPE funding for staff?
(h) Are further academic programmes available (e.g., MSc or PhD)?
(i) Is training provided when new equipment or procedures are introduced into the department?
(j) Who provides this training?
(k) Are there compulsory programmes on health and safety, and on radiation protection? If so, at what intervals?
(l) Is there an orientation programme for new staff?

Appendix I

RADIATION ONCOLOGY IN LIMITED RESOURCE SETTINGS

I.1. BACKGROUND

This appendix describes the essential components needed to start a radiotherapy clinic in a setting with limited resources and illustrates its natural progression to a centre of competence and eventually to a centre of excellence. The key to describing the operation of a radiation oncology facility is the need to consider its three principal components: equipment, human resources and procedures. It is obvious that in order to start operations, a facility must be equipped. However, the failure of a radiotherapy centre to operate efficiently is frequently caused by limited human resources: an insufficient number of staff or inadequately educated staff. Another common barrier to effective operation is a lack of sensible procedures based on an examination of treatment outcomes. In order to operate a radiotherapy centre effectively, efficiently and safely, it is necessary to have appropriate equipment, dedicated and properly trained staff, and sensible procedures geared to the economic situation in the region.

This publication does not seek to establish a universal standard in any of the three categories of equipment, human resources or procedures. In order to judge the level of a particular facility as being basic, competent or excellent would require an examination of the facility's operation by an expert panel, taking into account its economic environment. For example, a modestly equipped radiotherapy centre in a resource poor setting but with appropriate staffing and effective procedures might qualify as a centre of excellence. On the other hand, the limited equipment would relegate such a clinic to the status of a basic centre in a resource rich environment, where excellence is judged entirely on the ability of the centre to deal with very rare and special cases. It is important that managers and decision makers in resource poor settings realize that excellence is based on the clinic's procedures and how the centre utilizes its available resources rather than on the sophistication of its equipment.

The concept of sustainability is important in the process of analysing the level of operation of a particular facility. The term 'basic' implies only that the clinic has the essential equipment and adequate staffing to treat most tumours with the intent of achieving local control of the disease to the extent this is possible. In addition, their procedures must be reasonable and consistent with basic operation. A centre would be classified as 'competent' when, in order to ensure operation in the long term, it is able to provide clinical training to its entire staff on-site. Such a clinic would be able to educate its own RTTs completely and arrange for specialized academic training for its radiation oncologists and medical physicists at some other site. It would have adequate patient follow-up to track treatment outcomes, but there might not be a national cancer registry. The term 'excellence' would be applicable to a centre that serves the needs of other centres in the region by providing them with a site for clinical training and engaging in clinical research to improve the treatment outcomes for those tumours and stages of disease that are common to its region. In other words, a centre of competence is more or less self-sustainable, whereas a centre of excellence actually contributes to the sustainability of cancer treatment in the region it is located in and has a greater impact than a centre of competence.

Sophisticated equipment does not treat cancer by itself. Radiation oncologists, aided by other trained professionals, do.

I.2. THE BASIC RADIATION ONCOLOGY CLINIC

I.2.1. Recommended equipment and staffing levels

The term 'basic radiation oncology clinic' implies that the clinic has the essential equipment and adequate staffing to treat most tumours, with the intent of achieving local control of the disease to the extent possible. The clinic operates a cancer registry and has procedures for follow-up of treated patients.

Table I.1 lists the requirements for building, equipment and staffing that ought to be found in a basic cancer therapy centre that treats approximately 500 new patients per annum with teletherapy (with the intent of curing about 50% of them), and about 200 patients per annum with brachytherapy. The work is organized into two shifts. Staff needs should be adjusted to the number of patients treated. The training of staff requires that senior professionals or specialized trainers be available at the clinic.

Basic centres are equipped with a ^{60}Co unit or a single energy linac without a multileaf collimator, portal imaging or networking. With increasing complexity of radiotherapy treatment, for example, from a simple treatment with ^{60}Co using standard blocks to conformal radiotherapy with a multimode linac, the number of staff (especially physics staff) will need to increase.

I.2.2. Treatment procedures and clinic management

The equipment and staffing indicated above would be sufficient to start operations but certainly would not be sustainable without adding a training

TABLE I.1. ESSENTIAL EQUIPMENT AND STAFFING FOR A BASIC RADIOTHERAPY CLINIC

Set-up	Equipment/staffing
Building	One megavoltage bunker (desirable: space for one more) One X ray bunker for the orthovoltage unit A simulator room A dark room (for film processing) A dosimetry planning/physicist room (equipment storage necessary) One HDR bunker (or an LDR room)[a] A mould room Ample clinical space (with waiting, consulting, changing and examination rooms)
External beam therapy equipment	One single photon energy teletherapy unit One orthovoltage unit Beam measurement and QA + RP[b] physics equipment Simulator, preferably a CT simulator (otherwise desirable for access to the CT) A computerized TPS Film processing equipment Patient immobilization devices and mould room equipment
Brachytherapy HDR or LDR equipment[a]	One brachytherapy afterloader (two or more if LDR) An X ray C-arm for verification A computerized TPS (if LDR, it can be integrated into the external beam TPS) A full range of applicators Quality assurance physics equipment
Personnel	Four or five radiation oncologists[c] Three or four medical physics staff[d] Seven RTTs Three oncology nurses[c] One maintenance technician/engineer

[a] HDR versus LDR: An LDR brachytherapy unit can treat only approximately 100 patients per annum. Sites with a larger number of cervical cancer cases require HDR brachytherapy.

[b] RP: radiation protection.

[c] An increase by 50% is required if the staff are also responsible for chemotherapy; in this case a chemotherapy suite must be available.

[d] This requires at least one, and preferably two, senior clinically qualified radiotherapy medical physicist(s). Other physics staff must be clinically qualified radiotherapy medical physicists, resident physicists or dosimetrists.

component. Hence, to qualify as a centre of competence, the clinic should provide training to replace its own RTTs. In addition, it should be able to provide financial resources to enable academic training for replacement radiation oncologists and medical physicists, as well as on-site clinical training for these professionals. A centre of competence should practice and promote a culture of QA as evidenced by written policies and procedures guiding the treatment of its patients, and carry out regular preventive maintenance of its equipment. In addition, peer review of the clinical procedures, regular evaluation of morbidity and mortality (with special attention paid to unanticipated adverse events) as well as regular analysis of both short term and long term outcomes with regard to tumour control for the most common types of cancer are essential by regularly following up the treated patients.

In order to increase the level of impact from the local environment to the national or regional level, and thereby to qualify as a centre of excellence, the clinic should be a resource available to other centres for training. It should be investigating improved methods of therapy to treat the most common cancers in the region, hopefully contributing to the research literature and thereby providing guidance to other centres in the region. To do this, the centre should be associated with a cancer registry at least hospital based but ideally a national registry meeting the standards of the International Agency for Research on Cancer (IARC). In addition, the centre should engage in associated cancer control activities, such as cancer prevention (e.g., tobacco control as well as vaccination against human papillomavirus (HPV) and hepatitis B), early diagnosis (e.g., Pap smears) and palliative care (e.g., morphine for pain control).

I.3. TRANSITION FROM BASIC OPERATION TO COMPETENCE TO EXCELLENCE

The type, amount and level of sophistication of the equipment available do not determine the level of a centre's operation. Rather, its ability to operate self-sustainably through education and its ability to engage in analysis of its own treatment outcomes, thereby providing guidance for others and creating an impact in the region would be the defining characteristics. It is only when a centre is able to provide evidence demonstrating that it has achieved the status of at least a centre of competence and preferably a centre of excellence that managers should seek to introduce sophisticated or leading edge technologies, which require a much higher level of education and training of staff for implementation to be effective and sustainable.

Appendix II

REMARKS ON THE CONSISTENCY OF THE TERMINOLOGY USED IN RADIOTHERAPY

II.1. INTRODUCTION

In order to avoid misconceptions and misunderstandings in the use of terminology at various radiotherapy departments worldwide, auditors are encouraged to make themselves familiar with the explanations given in this appendix. These have been devised for the purpose of consistency. However, this does not constitute an intention to exclude other definitions of these various terms.

II.2. PATIENT

The patient is an individual with one or more cancers.

II.3. CANCER CASE

A cancer case is a new cancer registered, possibly several different cancers in a single individual (synchronous or metachronous cancers).

II.4. TREATMENT OR COURSE OF TREATMENT

A treatment is a course of radiotherapy consisting of a number of sessions, treating a given cancer. Whether the cancer is in one or several different target volumes (T and N), the treatment is still considered as one treatment. An additional irradiation at a distance from the primary (e.g. prophylactic cranial irradiation in small cell lung cancer) could be considered a different course of treatment, since the additional workload linked to it might amount to a new treatment (with a different simulation, a different set-up at the treatment machine and a different dose calculation).

The auditors should note in their report what comprises a treatment at the audited department and give some examples.

II.5. TREATMENT PLAN

A treatment plan involves at least a 2-D distribution of doses.

II.6. TREATMENT SESSION

A treatment session is synonymous with a fraction. One irradiation session comprises one or more fields on one or more target volumes for the same patient. Sessions are sometimes understood as a time slot at a treatment machine (ten minutes, for example). A complex treatment might use more than one time slot (e.g. a treatment of a child with medulloblastoma); therefore, it can be registered as one session or as several sessions depending on the departmental definition. Auditors need to clarify what is understood as a treatment session in an audited department, and the report of the audit must be unambiguous in this matter.

II.7. TREATMENT FIELD

A treatment field is a single radiation beam. Each beam orientation may include more than one field size. Auditors need to determine what definition is used.

II.8. SHIFT

A shift is normal working hours for a given professional class. A department might be open for longer daily hours and therefore use successive shifts for its personnel.

II.9. WORKLOAD

The workload of a radiotherapy department is determined by the number of treatments provided.

Appendix III

REMARKS ON THE ENUMERATION OF PATIENTS AND CANCER CASES

III.1. INTRODUCTION

While the concept of a 'patient' is uncontroversial, the number of 'cancer cases' is recorded and reported differently not only in developing countries but also in industrialized countries and from institution to institution. The auditors must establish the basis from which these statistics are derived.

III.2. CATCHMENT AREA

Are the cancer cases recorded an attempt to create a national or a regional cancer registry derived from the entire country or a region of the country?

Are the cancer cases derived from patients presenting to all the hospitals affiliated to the major hospital being audited or only from those presenting to the audited institution?

III.3. SOURCES OF INFORMATION

Do the cases include clinical and pathological diagnoses or only the latter?

III.4. MANAGEMENT

Do these cases include patients who may have simply been sent home for terminal care or those managed by surgery or chemotherapy in addition to those seen in a combined assessment clinic? Or are the cases recorded only those who have received radiotherapy?

III.5. SKIN CANCER: INCLUSIONS AND EXCLUSIONS

Do the cases recorded include all cases of skin cancer or only malignant melanomas (in conformity with IARC guidelines for national cancer registries?) Are all cases of Kaposi's sarcoma (AIDS and HIV negative) included?

III.6. COUNTING

It is usual to count a patient with a synchronous or metachronous cancer at a second primary site as a second case. In some institutions, the development of metastases subsequent to primary management is recorded as a further case.

ACKNOWLEDGEMENTS

The IAEA gratefully acknowledges major contributions to this publication by: Scalliet, P. (Belgium); Thwaites, D. (United Kingdom); Järvinen, H. (Finland); Coffey, M. (Ireland). The IAEA also wishes to acknowledge valuable suggestions and criticisms from the participants of the QUATRO workshop held in Vienna from 9 to 11 May 2005: Brunetto, M. (Argentina); Kron, T. (Australia); Smoke, M. (Canada); Cheung, K. (China); Castellanos, M.E. (Colombia); Alfonso, R. (Cuba); Novotny, J. (Czech Republic); Nyström, H. (Denmark); El-Gantiry, M. (Egypt); Salminen, E. (Finland); Kataria-Sethi, T. (India); Wadhawan, G.S. (India); Yusop, H.M.M. (Malaysia); El Gueddari, B. (Morocco); Ibn Seddik, A. (Morocco); Olusoji Ojebode, J. (Nigeria); Calaguas, M. (Philippines); Bulski, W. (Poland); Maciejewski, B. (Poland); Engel-Hills, P. (South Africa); Van der Merwe, D. (South Africa); Kunkler, I. (United Kingdom); Stewart-Lord, A. (United Kingdom); Acevedo, T. (Uruguay); Zubizaretta, E. (Uruguay).

REFERENCES

[1] KUTCHER, G.J., et al., Report of AAPM TG 40, Comprehensive QA for radiation oncology, Med. Phys. **21** (1994) 581–618.

[2] THWAITES, D.I., SCALLIET, P., LEER, J.W., OVERGAARD, J., Quality assurance in radiotherapy, Radiother. Oncol. **35** (1995) 61–73.

[3] LEER, J.W., McKENZIE, A., SCALLIET, P., THWAITES, D.I., Practical Guidelines for the Implementation of a Quality System in Radiotherapy, ESTRO Physics for Clinical Radiotherapy Booklet No. 4, European Society for Therapeutic Radiology and Oncology, Brussels (1998).

[4] IŻEWSKA, J., SVENSSON, H., IBBOTT, G., "Worldwide quality assurance networks for radiotherapy dosimetry", Standards and Codes of Practice in Medical Radiation Dosimetry (Proc. Int. Symp. Vienna, 2002), Vol. 2, IAEA, Vienna (2004) 139–155.

[5] IŻEWSKA, J., ANDREO, P., The IAEA/WHO TLD postal programme for radiotherapy hospitals, Radiother. Oncol. **54** (2000) 65–72.

[6] INTERNATIONAL ATOMIC ENERGY AGENCY, Standardized quality audit procedures for on-site dosimetry review visits to radiotherapy hospitals, SSDL Newsletter No. 46, IAEA, Vienna (2002) 17–23.

[7] INTERNATIONAL ATOMIC ENERGY AGENCY, On-site Visits to Radiotherapy Centres: Medical Physics Procedures, Quality Assurance Team for Radiation Oncology (QUATRO), IAEA-TECDOC-1543, IAEA, Vienna (in preparation)

[8] INTERNATIONAL ATOMIC ENERGY AGENCY, Setting up a Radiotherapy Programme: Clinical, Medical Physics, Radiation Protection and Safety Aspects, IAEA, Vienna (in preparation).

[9] INTERNATIONAL COMMISSION ON RADIATION UNITS AND MEASUREMENTS, Prescribing, Recording and Reporting Photon Beam Therapy, ICRU Rep. 50, ICRU, Bethesda, MD (1993).

[10] INTERNATIONAL COMMISSION ON RADIATION UNITS AND MEASUREMENTS, Prescribing, Recording and Reporting Photon Beam Therapy (Suppl. to ICRU Rep. 50), ICRU Rep. 62, ICRU, Bethesda, MD (1999).

[11] FOOD AND AGRICULTURE ORGANIZATION OF THE UNITED NATIONS, INTERNATIONAL ATOMIC ENERGY AGENCY, INTERNATIONAL LABOUR ORGANISATION, OECD NUCLEAR ENERGY AGENCY, PAN AMERICAN HEALTH ORGANIZATION, WORLD HEALTH ORGANIZATION, International Basic Safety Standards for Protection against Ionizing Radiation and for the Safety of Radiation Sources, Safety Series No. 115, IAEA, Vienna (1996).

[12] INTERNATIONAL ATOMIC ENERGY AGENCY, Absorbed Dose Determination in External Beam Radiotherapy: An International Code of Practice for Dosimetry Based on Standards of Absorbed Dose to Water, Technical Reports Series No. 398, IAEA, Vienna (2000).

[13] INTERNATIONAL ATOMIC ENERGY AGENCY, Calibration of Photon and Beta Ray Sources Used in Brachytherapy: Guidelines on Standardized Procedures at Secondary Standards Dosimetry Laboratories (SSDLs) and Hospitals, IAEA-TECDOC-1274, IAEA, Vienna (2002).

CONTRIBUTORS TO DRAFTING AND REVIEW

Abratt, R.	Groote Schuur Hospital, South Africa
Aguirre, F.	MD Anderson Cancer Center, United States of America
Andreo, P.	International Atomic Energy Agency
Coffey, M.	St. James's Hospital, Ireland
Drew, J.	University of Sydney, Australia
El Gueddari, B.	Institut National d'Oncologie, Morocco
Hanson, W.	MD Anderson Cancer Center, United States of America
Iżewska, J.	International Atomic Energy Agency
Järvinen, H.	Radiation and Nuclear Safety Authority (STUK), Finland
Kiel, K.	Northwestern Memorial Hospital, United States of America
Lartigau, E.	Centre Oscar Lambret, France
Leborgne, F.	Pereira Rossell Hospital, Uruguay
Levin, V.	International Atomic Energy Agency
McKenzie, A.	Bristol Haematology & Oncology Centre, United Kingdom
Meghzifene, A.	International Atomic Energy Agency
Poitevin, A.	Instituto Nacional de Cancerología, Mexico
Scalliet, P.	UCL Saint-Luc University Hospital, Belgium
Shortt, K.	International Atomic Energy Agency

Smyth, V.	National Radiation Laboratory, New Zealand
Svensson, H.	Karolinska Institute, Sweden
Thwaites, D.	Yorkshire Cancer Centre, United Kingdom
Vikram, B.	International Atomic Energy Agency

Consultants Meetings

Vienna, Austria: 17–21 February 2003, 16–20 February 2004

IAEA Workshop: Quality Assurance Team for Radiation Oncology (QUATRO)

Vienna, Austria, 9–11 May 2005

No. 21, July 2006

Where to order IAEA publications

In the following countries IAEA publications may be purchased from the sources listed below, or from major local booksellers. Payment may be made in local currency or with UNESCO coupons.

Australia
DA Information Services, 648 Whitehorse Road, Mitcham Victoria 3132
Telephone: +61 3 9210 7777 • Fax: +61 3 9210 7788
Email: service@dadirect.com.au • Web site: http://www.dadirect.com.au

Belgium
Jean de Lannoy, avenue du Roi 202, B-1190 Brussels
Telephone: +32 2 538 43 08 • Fax: +32 2 538 08 41
Email: jean.de.lannoy@infoboard.be • Web site: http://www.jean-de-lannoy.be

Canada
Bernan Associates, 4611-F Assembly Drive, Lanham, MD 20706-4391, USA
Telephone: 1-800-865-3457 • Fax: 1-800-865-3450
Email: order@bernan.com • Web site: http://www.bernan.com

Renouf Publishing Company Ltd., 1-5369 Canotek Rd., Ottawa, Ontario, K1J 9J3
Telephone: +613 745 2665 • Fax: +613 745 7660
Email: order.dept@renoufbooks.com • Web site: http://www.renoufbooks.com

China
IAEA Publications in Chinese: China Nuclear Energy Industry Corporation, Translation Section, P.O. Box 2103, Beijing

Czech Republic
Suweco CZ, S.R.O. Klecakova 347, 180 21 Praha 9
Telephone: +420 26603 5364 • Fax: +420 28482 1646
Email: nakup@suweco.cz • Web site: http://www.suweco.cz

Finland
Akateeminen Kirjakauppa, PL 128 (Keskuskatu 1), FIN-00101 Helsinki
Telephone: +358 9 121 41 • Fax: +358 9 121 4450
Email: akatilaus@akateeminen.com • Web site: http://www.akateeminen.com

France
Form-Edit, 5, rue Janssen, P.O. Box 25, F-75921 Paris Cedex 19
Telephone: +33 1 42 01 49 49 • Fax: +33 1 42 01 90 90 • Email: formedit@formedit.fr

Lavoisier SAS, 14 rue de Provigny, 94236 Cachan Cedex
Telephone: +33 1 47 40 67 00 • Fax +33 1 47 40 67 02
Email: livres@lavoisier.fr • Web site: http://www.lavoisier.fr

Germany
UNO-Verlag, Vertriebs- und Verlags GmbH, August-Bebel-Allee 6, D-53175 Bonn
Telephone: +49 02 28 949 02-0 • Fax: +49 02 28 949 02-22
Email: info@uno-verlag.de • Web site: http://www.uno-verlag.de

Hungary
Librotrade Ltd., Book Import, P.O. Box 126, H-1656 Budapest
Telephone: +36 1 257 7777 • Fax: +36 1 257 7472 • Email: books@librotrade.hu

India
Allied Publishers Group, 1st Floor, Dubash House, 15, J. N. Heredia Marg, Ballard Estate, Mumbai 400 001,
Telephone: +91 22 22617926/27 • Fax: +91 22 22617928
Email: alliedpl@vsnl.com • Web site: http://www.alliedpublishers.com

Bookwell, 2/72, Nirankari Colony, Delhi 110009
Telephone: +91 11 23268786, +91 11 23257264 • Fax: +91 11 23281315
Email: bookwell@vsnl.net

Italy
Libreria Scientifica Dott. Lucio di Biasio "AEIOU", Via Coronelli 6, I-20146 Milan
Telephone: +39 02 48 95 45 52 or 48 95 45 62 • Fax: +39 02 48 95 45 48

Japan
Maruzen Company, Ltd., 13-6 Nihonbashi, 3 chome, Chuo-ku, Tokyo 103-0027
Telephone: +81 3 3275 8582 • Fax: +81 3 3275 9072
Email: journal@maruzen.co.jp • Web site: http://www.maruzen.co.jp

Korea, Republic of
KINS Inc., Information Business Dept. Samho Bldg. 2nd Floor, 275-1 Yang Jae-dong SeoCho-G, Seoul 137-130
Telephone: +02 589 1740 • Fax: +02 589 1746
Email: sj8142@kins.co.kr • Web site: http://www.kins.co.kr

Netherlands
De Lindeboom Internationale Publicaties B.V., M.A. de Ruyterstraat 20A, NL-7482 BZ Haaksbergen
Telephone: +31 (0) 53 5740004 • Fax: +31 (0) 53 5729296
Email: books@delindeboom.com • Web site: http://www.delindeboom.com

Martinus Nijhoff International, Koraalrood 50, P.O. Box 1853, 2700 CZ Zoetermeer
Telephone: +31 793 684 400 • Fax: +31 793 615 698 • Email: info@nijhoff.nl • Web site: http://www.nijhoff.nl

Swets and Zeitlinger b.v., P.O. Box 830, 2160 SZ Lisse
Telephone: +31 252 435 111 • Fax: +31 252 415 888 • Email: infoho@swets.nl • Web site: http://www.swets.nl

New Zealand
DA Information Services, 648 Whitehorse Road, MITCHAM 3132, Australia
Telephone: +61 3 9210 7777 • Fax: +61 3 9210 7788
Email: service@dadirect.com.au • Web site: http://www.dadirect.com.au

Slovenia
Cankarjeva Zalozba d.d., Kopitarjeva 2, SI-1512 Ljubljana
Telephone: +386 1 432 31 44 • Fax: +386 1 230 14 35
Email: import.books@cankarjeva-z.si • Web site: http://www.cankarjeva-z.si/uvoz

Spain
Díaz de Santos, S.A., c/ Juan Bravo, 3A, E-28006 Madrid
Telephone: +34 91 781 94 80 • Fax: +34 91 575 55 63 • Email: compras@diazdesantos.es
carmela@diazdesantos.es • barcelona@diazdesantos.es • julio@diazdesantos.es
Web site: http://www.diazdesantos.es

United Kingdom
The Stationery Office Ltd, International Sales Agency, PO Box 29, Norwich, NR3 1 GN
Telephone (orders): +44 870 600 5552 • (enquiries): +44 207 873 8372 • Fax: +44 207 873 8203
Email (orders): book.orders@tso.co.uk • (enquiries): book.enquiries@tso.co.uk • Web site: http://www.tso.co.uk

On-line orders:
DELTA Int. Book Wholesalers Ltd., 39 Alexandra Road, Addlestone, Surrey, KT15 2PQ
Email: info@profbooks.com • Web site: http://www.profbooks.com

Books on the Environment:
Earthprint Ltd., P.O. Box 119, Stevenage SG1 4TP
Telephone: +44 1438748111 • Fax: +44 1438748844
Email: orders@earthprint.com • Web site: http://www.earthprint.com

United Nations (UN)
Dept. I004, Room DC2-0853, First Avenue at 46th Street, New York, N.Y. 10017, USA
Telephone: +800 253-9646 or +212 963-8302 • Fax: +212 963-3489
Email: publications@un.org • Web site: http://www.un.org

United States of America
Bernan Associates, 4611-F Assembly Drive, Lanham, MD 20706-4391
Telephone: 1-800-865-3457 • Fax: 1-800-865-3450
Email: order@bernan.com • Web site: http://www.bernan.com

Renouf Publishing Company Ltd., 812 Proctor Ave., Ogdensburg, NY, 13669
Telephone: +888 551 7470 (toll-free) • Fax: +888 568 8546 (toll-free)
Email: order.dept@renoufbooks.com • Web site: http://www.renoufbooks.com

Orders and requests for information may also be addressed directly to:

Sales and Promotion Unit, International Atomic Energy Agency
Wagramer Strasse 5, P.O. Box 100, A-1400 Vienna, Austria
Telephone: +43 1 2600 22529 (or 22530) • Fax: +43 1 2600 29302
Email: sales.publications@iaea.org • Web site: http://www.iaea.org/books

07-13411